Agnes Giberne

Sun, Moon, and Stars

Astronomy for beginners

Agnes Giberne

Sun, Moon, and Stars
Astronomy for beginners

ISBN/EAN: 9783337412104

Printed in Europe, USA, Canada, Australia, Japan

Cover: Foto ©berggeist007 / pixelio.de

More available books at **www.hansebooks.com**

SUN, MOON, AND STARS.

ASTRONOMY FOR BEGINNERS.

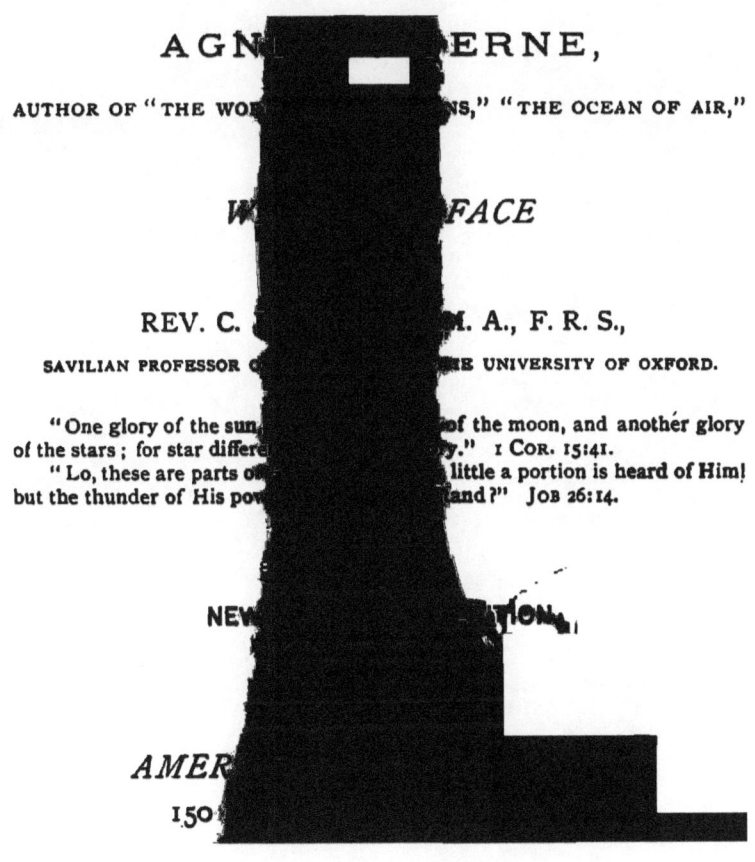

AGNES M. CLERKE,

AUTHOR OF "THE WORLD OF SUNS," "THE OCEAN OF AIR,"

WITH PREFACE

REV. C. PRITCHARD, M. A., F. R. S.,

SAVILIAN PROFESSOR OF ASTRONOMY, UNIVERSITY OF OXFORD.

"One glory of the sun, and another of the moon, and another glory of the stars; for star differeth from star in glory." 1 COR. 15:41.
"Lo, these are parts of His ways: but how little a portion is heard of Him! but the thunder of His power who can understand?" JOB 26:14.

NEW EDITION.

AMERICAN...
150...

PREFACE TO REVISED EDITION.

IN the course of even a few years changes of necessity take place in so progressive a science as Astronomy. Some new discoveries are made, some new ideas are started, some old beliefs have to be given up. When the twentieth thousand of the English edition of "Sun, Moon, and Stars" was called for, it seemed to me that the time had come for a thorough revision of the whole: for a bringing up of the information contained therein, as far as possible, up to date.

In the carrying out of this aim no trouble has been spared, and able friends have generously given time to helping me. Many old passages have been omitted and many new ones interpolated, large portions of chapters have been rewritten, and the two last chapters are almost fresh ones. My little book will, I hope, go forth replenished with new vigor for a new campaign.

AGNES GIBERNE.

WORTON HOUSE,
    EASTBOURNE, December, 1892.

PREFACE.

THE pages of this little book on a great subject were submitted to my criticism while passing through the press, with a request from a friend that I would make any suggestions which might occur to me for its improvement. Naturally such a request was entertained in the first instance with hesitation and misgiving. But after a rapid perusal of the first sheet I found my interest awakened and then gradually secured; for the book seemed to me to possess certain features of no ordinary character, and, in my judgment, held out the promise of supplying an undoubted want, thus enabling me to answer a question which I have been often asked, and which had as often puzzled me, to the effect, "*Can you tell me of any little book on Astronomy suited to beginners?*" I think that just such a book is here presented to the reader. For the tale of the Stellar Universe is therein told with great simplicity and perhaps with sufficient completeness, in an earnest and pleasant style, equally free, I think, from any considerable inaccuracy or any unpardonable exaggeration. We have here the outlines of elementary astronomy, not merely detailed without mathematics, but to a very great extent expressed in untechnical language. Success in such an attempt I am inclined to regard as a considerable feat, and one of much practical utility.

For the science of astronomy is essentially a science of great magnitude and great difficulty. From the time of Hipparchus, some century and a half before the Christian era, down to the present day, its cultivation has severely taxed the minds of a succession of men endowed with the rarest genius. The facts and the truths of the science thus secured have been of very slow accretion; but like all other truths, when once secured and thoroughly understood, they are found to admit of very simple verbal expression, and to lie well within the general comprehension, and perhaps I may safely add within the sympathies of all educated men and women.

Thus the great astronomers, the original discoverers of the last twenty centuries, have labored each in his separate field of the vast universe of nature, and other men, endowed with other gifts, have entered on their labors, and by systematizing, correlating, and simplifying the expression of their results, have brought the whole within the grasp of cultivated men engaged in other branches of the varied pursuits of our complicated life. It is in this sort of order that the amelioration and civilization of mankind have proceeded, and at the present moment are, I hope and believe, rapidly proceeding.

It was, I suspect, under this point of view, though half unconsciously so, that my attention was arrested by the little book now presented to the reader. For we have here many of the chief results of the laborious researches of such men as Ptolemy, Kepler, Newton, Herschel, Fraunhofer, Janssen, Lockyer, Schiaparelli, and others—no matter where accumulated or by whom recorded—filtered through the mind of a thoughtful and cultured lady, and here presented to other minds in the very forms wherein they have been assimilated and pictured in her own. And the forms and pictures in the main are true. It is in this way that the intellectual "protoplasm" of the human mind is fostered and practically disseminated.

And then there is still another point of view from which this general dissemination of great truths in a simple style assumes an aspect of practical importance. I allude to the influences of this process on the imaginative or poetic side of our complex nature, and in support of the remark I shall quote from the pages of a work recently published by the Professor of Poetry in the University of Oxford: "Every new province of knowledge which science conquers, poetry may in turn enter into and possess. But this can only be done gradually. Before imagination can take up and mould the results of science these must cease to be difficult, laborious, abstruse. The knowledge of them must have become to the poet himself, and in some measure to his audience, familiar, habitual, spontaneous. And here we see how finely science and poetry may interact and minister to each other . . . . Some beginning of such a reconciling process we may see here and there in those poems of '*In Memoriam*,' in which the Poet Laureate has finely inwrought new truths of science into the texture of yearning affection and spiritual meditation . . . ."*

Wordsworth in one of his prefaces has stated so clearly the truth on this subject that I cannot do better than give his words. "If the time should ever come," he says, "when what is now science becomes familiarized to men, then the remotest discoveries of the chemist, the botanist, the mineralogist, will be as proper objects of the poet's art as any upon which it can be employed. He will be ready to follow the steps of the man of science, he will be at his side, carrying sensation into the midst of the objects of science itself. The poet will lend his divine spirit to aid the

* "On Poetic Interpretation of Nature," by J. C. Shairp, LL. D., Edinburgh. David Douglas, 1877.

transfiguration and will welcome the being thus produced as a dear and genuine inmate of the household of man."

It is for reasons such as those above stated that I heartily commend this little book to the attention of those of my countrymen and countrywomen who take an interest in the advancement of the intellectual progress and culture of society. The story of the Kosmos is told by the authoress in her own language, after her own method, and with no guidance of mine: in the main, I believe, as I have said, the story is correct; but I must disavow the responsibility of every detail.

C. PRITCHARD.

OXFORD, September 2, 1879.

AUTHOR'S PREFACE.

LITTLE remains to me, as needing to be said about my little book, after the most kind words of Prof. Pritchard.

For years past I have had it in my mind to endeavor to supply some day, so far as might be, that of which there appeared to be a serious want—a plain and easily-to-be-understood Introduction to Astronomy. My wish has been to make it, not merely a text-book for school use, though possibly it may serve for that also, but a volume of sufficient interest for general reading; not merely a book adapted for intelligent boys and girls, though I hope it may be found serviceable for them also, but a volume fitted for "beginners" of all kinds, whether children, working-men, or even grown people of the educated classes, who should have a desire to enter for the first time on the study of this fascinating science. The attempt, such as it is, has been at last accomplished.

I trust that a second underlying object or thought may not be found to have been altogether fruitless. In the Book of Nature, side by side with the Book of Revelation, we may learn some things about our Father in heaven. It would be happiness to feel that I had helped any, however slightly, to look upward through nature unto nature's God.

I have, lastly, to express my warm gratitude to Prof. Pritchard for the time and trouble he has so generously expended in looking through the proof-sheets of a work by one who, as a perfect stranger, had and could have no manner of claim upon him. A request made through a mutual friend, with doubt and uncertainty on my part, has met with a fulness of response and kindness utterly unexpected, and for which no words of thanks seem half sufficient.

EASTBOURNE, September 5, 1879.

CONTENTS.

PART I.

I. THE EARTH ONE OF A FAMILY	17
II. THE HEAD OF OUR FAMILY	28
III. WHAT BINDS THE FAMILY TOGETHER?	41
IV. THE LEADING MEMBERS OF OUR FAMILY.—FIRST GROUP	55
V. THE LEADING MEMBERS OF OUR FAMILY.—SECOND GROUP	65
VI. OUR PARTICULAR FRIEND AND ATTENDANT	76
VII. VISITORS	88
VIII. LITTLE SERVANTS	96
IX. NEIGHBORING FAMILIES	105
X. OUR NEIGHBORS' MOVEMENTS	118

PART II.

I. MORE ABOUT THE SOLAR SYSTEM	135
II. MORE ABOUT THE SUN	149
III. YET MORE ABOUT THE SUN	162
IV. MORE ABOUT THE MOON	172
V. YET MORE ABOUT THE MOON	182
VI. MERCURY, VENUS, AND MARS	194

Contents.

VII. JUPITER	209
VIII. SATURN	219
IX. URANUS AND NEPTUNE	227
X. COMETS AND METEORS	236
XI. MORE ABOUT COMETS AND METEORS	244

PART III.

I. MANY SUNS	257
II. SOME PARTICULAR SUNS	266
III. DIFFERENT KINDS OF SUNS	278
IV. GROUPS AND CLUSTERS OF SUNS	287
V. THE MILKY WAY	296
VI. READING THE LIGHT	307
VII. FURTHER THOUGHTS	317
TABLE OF SUBJECTS	329

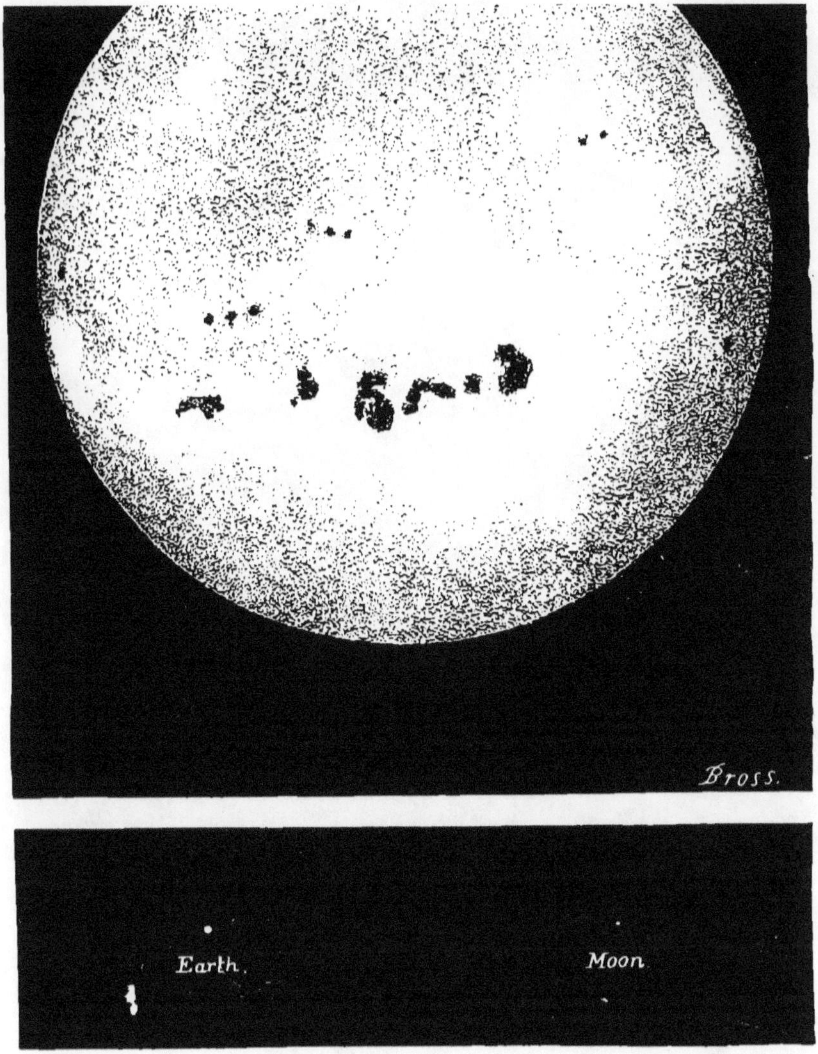

THE SUN, WITH COMPARATIVE MAGNITUDES OF THE EARTH AND MOON.

PART I.

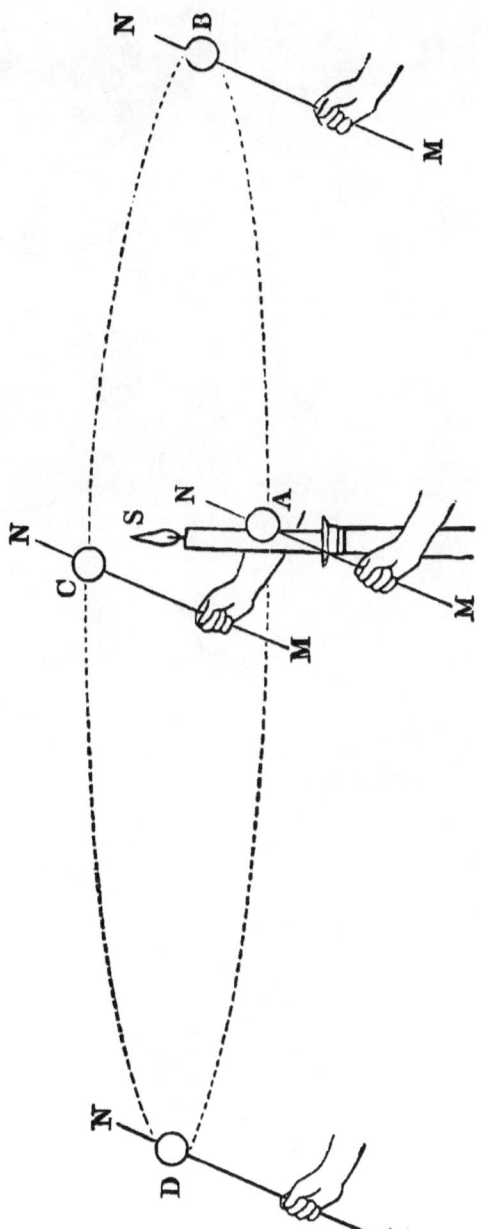

DIAGRAM ILLUSTRATING THE ORBIT OF THE EARTH ROUND THE SUN.

A B C D, the Ecliptic (a very round ellipse), here shown in perspective. S, the Sun, in one of the foci. A, the position of the Earth at the Autumnal Equinox. B, at the Winter Solstice. C, at the Vernal Equinox. D, at the Summer Solstice. M N, the direction of the Terrestrial Axis, preserving its direction, or parallelism, during the Earth's motion in her orbit, and always pointing towards stars in the constellations called Northern.

Sun, Moon, and Stars.

CHAPTER I.

THE EARTH ONE OF A FAMILY.

"The heaven, even the heavens, are the LORD'S: but the earth hath He given to the children of men."—PSA. 115:16.

WHAT IS this world of ours?

Something very great—and yet something very little. Something very great compared with the things upon the earth; something very little compared with the things outside the earth.

And as our journeyings together are for a while to be away from earth, we shall find ourselves obliged to count her as something quite small in the great universe, where so many larger and mightier things are to be found—if indeed they are mightier.

Not that we have to say good-by altogether to our old home. We must linger about her for a while before starting, and afterwards it will be often needful to come back, with speed swifter than the flight of light, that we may compare notes on the sizes and conditions of other places visited by us.

But first of all: what *is* this earth of ours?

It was rather a pleasant notion which men held in olden days, that we—that great and important "We" which loves to perch itself upon a height—stood firm and fixed at the very centre of everything.

The earth was supposed to be a vast flat plain, reaching nobody could tell how far. The sun rose and set for us alone; and the thousands of stars twinkled in the sky at night for nobody's good except ours; and the blue sky overhead was a crystal covering for the men of our earth, and nothing more. In fact, people seem to have counted themselves not merely to have had a kind of kingship over the lower animals of our earth, but to have been kings over the whole universe. Sun, stars, and sky, as well as earth, were made for man, and for man only.

This was the common belief, though even in those olden days there were *some* who knew better. But the world in general knows better now.

Earth the centre of the universe! Why, she is not that of even the particular family in the heavens to which she belongs. For we do not stand alone. The earth is one of a family of worlds, and that family is called THE SOLAR SYSTEM. And so far is our earth from being the head of the family that she is not even one of the more important members. She is merely one of the little sisters as it were.

Men not only believed the earth, in past days, to be at the centre of the universe, but also they believed her to remain there without change. Sun, moon, stars, planets, sky, might move, but never the earth. The solid ground beneath their feet, *that* at least was firm. Every day the sun rose and set, and every night the stars in like manner rose and set. But this was easily explained. We on our great earth stood firm and still, while sun, moon, stars, went circling round us once in every twenty-four hours, just for our sole and particular convenience. What an important personage man must have felt himself then in the great Universe!

Once again, we know better now!

For it is the earth that moves, and not the sun; it is the earth that moves, and not the stars. The daily movements of sun and stars, rising in the east, travelling over the sky and setting in the west, are no more real movements on their part than, when we travel in a railway-train, the seeming rush of hedges, telegraph-posts, houses, and fields is real. They are fixed and we are moving, yet the movement appears to us to be not ours but theirs.

Still more strongly would this appear to be the case if there were no noise, no shaking, no jarring and trembling, to make us feel that we are not at rest. Sometimes when a train begins to move gently out of a station, from among other trains, it is at the first moment quite impossible to say whether the movement belongs to the train in which we are seated or to a neighboring train. And in the motions of our earth there is no noise, no shaking, no jarring—all are rapid, silent, and even.

If you were rising through the air in a balloon, you would at first only know your own movement by seeing the earth seem to drop away from beneath you. And just so we can only know the earth's movements by seeing how worlds around us *seem* to move in consequence.

Speaking now of the Universe, I mean thereby the whole material creation as far as the stars reach. Sometimes the word is used in this sense, and sometimes it is used only for a particular part of creation nearer to us than other parts. At present, however, we will put aside all thought of the second narrower meaning.

The wisest astronomer living cannot tell us how far the stars reach. We know now that there is no firm crystal covering over our heads, dotted with bright points here and there; but only the wide open sky or heaven, containing millions of stars, some nearer, some farther, some bright enough to be seen by us all, some only visible through a telescope.

People talk often of the stars being "set in space;" and the meaning of "space" is simply "room." Where you are must be space, or you could not be there. But it is when we get away from earth, and travel in thought through the wide fields of space, amid innumerable stars, that we begin to feel how vast it is, and what specks we are ourselves—nay, what a speck our very earth is, in this great and boundless creation.

For there is no getting to the borders of space. As one telescope after another is made, each one stronger in power and able to reach farther than the last, still more and more stars are seen, and yet more and more behind and beyond, in countless millions.

It is the same all round the earth. The old notion about our world being a flat plain has been long since given up. We know her now to be a round globe, not fixed, but floating like the stars in space. We find this wonderfully described in the Bible, long before men knew or could know what the words meant: " He stretcheth out the north over the empty place, and HANGETH THE EARTH UPON NOTHING."*

When you look *up* into the sky from England, you are looking exactly in the opposite direction from where you would be looking *up* if you were in Australia. For Australia's "up" is England's "down," and England's "down" is Australia's "up."

Or, to put it more truly, "up" is always in the direction straight away from this earth, on whatever part of it you may be standing; and "down" is always towards the centre of the earth.

All round the globe, in north, south, east, west, whether you are in Europe, Asia, Africa, or America, though you will see different stars in certain different quarters of the earth, still overhead you will find shining countless points of light.

*Job 26:7.

And now, what are these stars?

This is a matter on which people are often confused, and on which it is well to be quite clear before going one step farther.

Some of the stars you have most likely often noticed. The seven chief stars of the Great Bear are known to a large number of people; and there are few who have not admired the splendid constellation of Orion. Perhaps you also know the W-shape of Cassiopeia, and the brilliant shining of Sirius, and the soft glimmer of the Pleiades.

The different constellations or groups of principal stars have been watched by men for long ages past. They are called the fixed stars, for they do not change. How many thousands of years ago they were first arranged by men into these groups, and who first gave them their names, we cannot tell.

True, night by night, through century after century, they rise, and cross the sky, and set. But those are only seeming movements. Precisely as the turning round of the earth upon her axis, once in every twenty-four hours, makes the sun appear to rise and cross the sky and set in the day-time, so also the same turning of the earth makes the stars appear to do the same in the night-time.

There is another seeming movement among the stars which is only in seeming. Some come into view in summer which cannot be seen in winter; and some come into view in winter which cannot be seen in summer. For the sun, moving on his pathway through the sky, hides those stars which shine with him in the day-time. And as he passes from point to point of his pathway through the constellations, he conceals from us fresh groups of stars by day and allows fresh groups to appear by night. Speaking generally, however, the stars remain the same year after year, century after century. The groups may still be seen as of old, fixed and unchanging.

What are these stars?

Stars and planets have been spoken about. There is a great difference between the two.

Perhaps if you were asked whether the sun is most like to a star or a planet, you would be rather at a loss. And many who have admired the brilliant evening star, Venus, often to be seen after sunset, would be surprised to learn that the evening star is in reality no star at all.

A star is a sun. Our sun is nothing more nor less than a star. Each one of the so-called "fixed stars" that you see shining at night in the sky is a sun like our sun; only some of the stars are larger suns and some are smaller suns than ours.

The main reason why our sun looks so much larger and brighter than the stars is that he is so very much nearer to us. The stars are one and all at enormous distances from the earth. By-and-by we will go more closely into the matter of their distance compared with the distance of the sun.

At present it is enough to say that if many of the stars were placed just as near to us as our sun is placed they would look just as large and bright, while there are some that would look a great deal larger and brighter. And if our sun were to travel away from us to the distance of the very nearest of the little twinkling stars, he would dwindle down and down in size and brilliancy, till at last we should not be able to tell him apart from the rest of the stars.

I have told you that the stars are called "fixed" because they apparently do not change from age to age. Though the movement of the earth makes them seem all to sweep past every night in company, yet they do not travel in and out among one another, or backwards and forwards, or from side to side. At all events, if there be such changes, they are so slow and so small as to be exceedingly difficult to find out. Each group of stars keeps its own old shape, as for hundreds of years back.

But among these fixed stars there are certain stars which do go to and fro and backwards and forwards. Now they are to be seen in the middle of one constellation and now in the middle of another. These restless stars were long a great perplexity. Men named them Planets or Wanderers.

We know now that the planets are in reality not stars at all: and also that they are not nearly so far away from us as the fixed stars. In fact, they are simply members of our own family, the Solar System. They are worlds, more or less like the world we live in; and they travel round and round the sun as we do, each more or less near to him; and they depend upon him for heat and light in more or less the same manner as ourselves.

Therefore, just as our sun is a star, and stars are suns, so our earth or world is a planet, and planets are worlds. EARTH is the name we give to that particular world or planet on which we live.

Planets may generally be known from stars by the fact that they do not twinkle.

But the great difference between the two lies in the fact that a star shines by his own radiant dazzling light; whereas a planet shines by light reflected or borrowed from the sun.

CHAPTER II.

THE HEAD OF OUR FAMILY.

"The sun, which . . . rejoiceth as a strong man to run a race."—PSA. 19:4, 5.

PEOPLE began, very early in the history of the world, to pay close attention to the sun. And no wonder. We owe so much to his heat and light that the marvel would be if men had not thought much about him.

Was the sun really any larger than he looked, and if so, how much larger was he? And what was his distance from the earth?

These were two of the questions which puzzled our ancestors the longest. If once they could have settled exactly how far off the sun really was, they could easily have calculated his exact size; but this was just what they could not do.

So one man supposed that the sun must be quite near, and very little larger than he looked. Another thought he might be seventy-five miles in diameter. A third ventured to believe that he was larger than the country of Greece. A fourth was so bold as to imagine that he might even outweigh the earth herself!

After a while many attempts were made to measure the distance of the sun; and a great many different answers to this difficult question were given by different men, most of them very wide of the mark.

It is only of late years that the matter has been clearly settled. And indeed it was found quite lately that a mistake of no less than three millions of miles had been made, notwithstanding all the care and all the attention given. But though three millions of miles sounds a great deal, yet it is really very little—only a tiny portion of the whole.

For the distance of the sun from the earth is no less than about NINETY-TWO MILLIONS OF MILES.

Ninety-two millions of miles! Can you picture that to yourself?

Try to think what is meant by a thousand miles. Our earth is eight thousand miles in diameter. In other words, if you were to thrust a gigantic knitting-needle through her body, from the North Pole to the South Pole, it would have to be about eight thousand miles long.

To reach the thought of one million, you must picture *one thousand times one thousand.* Our earth is about twenty-five thousand miles round. If you were to start from the mouth of the River Amazon in South America, and journey straight round the whole earth on the equator till you came back to the same point, you would have travelled about twenty-five thousand miles.

But that would be a long way from a million miles. You would have gone only once round the earth. Now a rope one million miles in length could be wrapped, not once only, but *forty times*, round and round the earth.

And when you have managed to reach up to the thought of one million miles, you have then to remember that the sun's distance is ninety-two times as much again. So, to picture clearly to ourselves the actual meaning of " ninety-two millions of miles" is not easy.

Suppose it were possible to lay a railroad from here to the sun. If you could journey thither in a perfectly straight line, at the rate of thirty miles an hour, never pausing for one single minute night or day, you would reach the sun in just about three hundred and fifty years.

Thirty miles an hour is a slow train. Suppose

we double the speed, and make it an express train, rushing along at the pace of sixty miles an hour. Then you might hope to reach the end of your journey in one hundred and seventy-five years. If you had quitted this earth early in the reign of King George I., never stopping on your way, you would be just now, in the year 1893, arriving at the sun.

So much for the sun's distance from us. Now as to his size.

I have already mentioned that our earth's diameter—that is, her *through measure*, as, for instance, the line drawn straight from England through her centre to New Zealand—is about eight thousand miles.

This sounds a good deal. But what do you think of the diameter of the sun being no less than *eight hundred and sixty-five thousand miles?*

The one is eight thousand miles, the other is eight hundred thousand miles!

Suppose you had a long slender pole which would pass through the middle of the earth, one end just showing at the North Pole and the other at the South Pole. You would need more than a hundred and eight of such poles, all joined together, to show the diameter of the sun.

The sun seems not to be made of nearly such heavy materials as the earth. He is what astronomers call less "dense," less close and compact in his make, just as wood is less dense and heavy than iron.

Still his size is so enormous that if you could have a pair of gigantic scales, and put the sun into one scale and the earth with every one of her brother and sister planets into the other, the sun's side would go down like lightning. He would be found to weigh seven hundred and fifty times as much as all the rest put together.

It would take more than twelve hundred thousands of little earths like ours, rolled into one huge ball, to make a globe as large as the sun.

Suppose our earth were to dwindle down and down, smaller and smaller, till she became a tiny globe only one inch in diameter. Imagine the sun at the same time dwindling down in the same manner, and keeping the same proportion as to size. "Keeping the same proportion" means that he would still be twelve hundred thousand times as large as the earth, that his axis, or the straight line through his centre from North Pole to South

Pole, would still be one hundred and eight times as long as the earth's axis.

Then, side by side with the minute ball, one inch in diameter, you would see a great globe nine feet in diameter—half as high again as a tall man.

And if you wish to gain a fair notion, not only of the sun's size, but of his distance from us, set such a nine-feet globe in a field, and move slowly round it a one-inch ball at a distance of three hundred and twenty yards. This may help you to understand.

In the beginning of the seventeenth century a man named Fabricius was startled by the sight of a certain black spot upon the face of the sun.

He watched till too dazzled to look any longer, supposing it to be a small cloud, yet anxious to learn more. Next day the spot was there still, but it seemed to have moved on a little way. Morning after morning this movement was found to continue, and soon a second spot, and then a third spot, were observed creeping in like manner across the sun. After a while they vanished, one at a time, round his edge, as it were; but after some days of patient waiting on the part of the lookers-on, they appeared again at the opposite edge and once more began their journey across.

Fabricius seems to have been the first, but he was not the last, to watch sun-spots. Many astronomers have given close attention to them. Modern telescopes, and the modern plan of looking at the sun through darkened glass, have made this possible in a way that was not possible two or three hundred years ago.

The first important discovery made through the spots on the sun was that the sun turns round upon his axis just in the same manner that the earth turns round upon hers. Instead of doing so once in the course of each twenty-four hours, like the earth, he turns once in the course of about twenty-five days.

It must not be supposed that the spots seen now upon the sun are the same spots that Fabricius saw in the reign of our James I. There is perpetual change going on—new spots forming, old spots vanishing, one spot breaking into two, two spots joining into one, and so on. Even in a single hour great alterations are sometimes seen to take place.

Still many of the spots do remain long enough and keep their shapes closely enough to be watched from day to day, and to be known again as old friends when they reappear after being about twelve days hidden on the other side of the sun. So that the turning of the sun upon his axis has become, after long and careful examination, a certain fact.

For more than thirty years one astronomer kept close watch over the spots on every day that it was possible to see the sun. Much was learned from his perseverance, and far more has been learned since.

Two or three months may be counted a fair medium length of time for a spot to endure, while one has been known to exist as long as eighteen months. This, however, was exceptional.

It seems pretty sure that these spots are caused by some kind of cyclones, great in extent and violence, on the surface of the sun.

We do not yet know with absolute certainty whether the sun is through and through one vast mass of heated gases; but this is the opinion now generally held; and the older idea of a possibly solid and even cool interior is at present given up.

The round, shining disc or flat surface seen by all of us is called the "photosphere" or "light-sphere." It has a tolerably well-defined edge or "limb," and dazzles the eyes with its intense radiance.

Of the photosphere Prof. Young writes, "All that we can learn as to the temperature and constitution of the sun makes it hardly less than certain that the visible surface, which is called the photosphere, is just a sheet of self-luminous cloud, precisely like the clouds of our own atmosphere, with the exception that the droplets of water, which constitute terrestrial clouds, are replaced in the sun by drops of molten metal and that the solar atmosphere in which they float is the flame of a burning, fiery furnace, raging with a fury and an intensity beyond all human conception." And again, "The photosphere is a shell of luminous clouds, formed by the cooling and condensation of the condensible vapors at the surface where exposed to the cold of outer space." Of course the words "cooling" and "cold" are comparative terms only, in reference to such a neighborhood.

Outside the photosphere lies the *chromosphere* or *chromatosphere* or *sierra*. This, seen at the edge of the sun, against the sky, has been described as "a quivering fringe of fire;" and since the sun turns incessantly upon his axis, presenting each hour fresh portions of his surface in that position, it follows that the whole surface of the sun is one restless, billowy ocean of fire. The waves of the sea on a stormy day, perceived in the far distance rising and breaking the horizon-line, may serve as a tiny illustration, only that in the sun the billows are of fire, not water. Think what their height must be that they should be thus visible—even through a telescope—at a distance of nearly ninety-three millions of miles!

It is believed that the chromosphere is formed of such gases—chiefly hydrogen—as refuse to be condensed into the molten clouds which form the photosphere.

Sun, as well as earth, possesses an atmosphere, but it is an atmosphere very different in description; not fitted like ours to support life, not "shallow and quiet," but "an envelope of matter, partly gaseous and partly perhaps pulverulent or smoke-like, many thousand miles in depth, and always most profoundly and violently agitated."*

Outside the chromosphere are to be seen bright, rose-colored "prominences"—extraordinary protuberant shapes of enormous size, though dwindled down to smallness by distance. During an eclipse, when the dark body of the moon glides between

* Prof. Young.

the sun and us, exactly covering the photosphere and hiding its glare, these crimson jagged tips stand out distinctly beyond the edge of sun and moon. Once upon a time they could be seen only at such seasons, during the brief space of a total eclipse; but the spectroscope now enables astronomers to see them in full daylight. Careful watching for hours together has thus become possible, not to speak of photographing.

They are found to be prolongations of the red chromosphere — that sea of glowing gas which bathes the whole body of the sun. They too consist of gas, mainly of hydrogen gas, rising like enormous mountain-billows out of that fiery ocean. "Flames" they are often called, though not consuming flames, like those which burn away substances in our earthly atmosphere, but rather continuously glowing masses of gas, luminous with vehement heat.

The height of these prominences has been repeatedly measured. Some are so lofty that ten little earths such as ours might be heaped up, one upon another, without reaching the top. One was seen to tower to the extent of 300,000 miles— an outrush of radiant gas from the sun almost incredible!

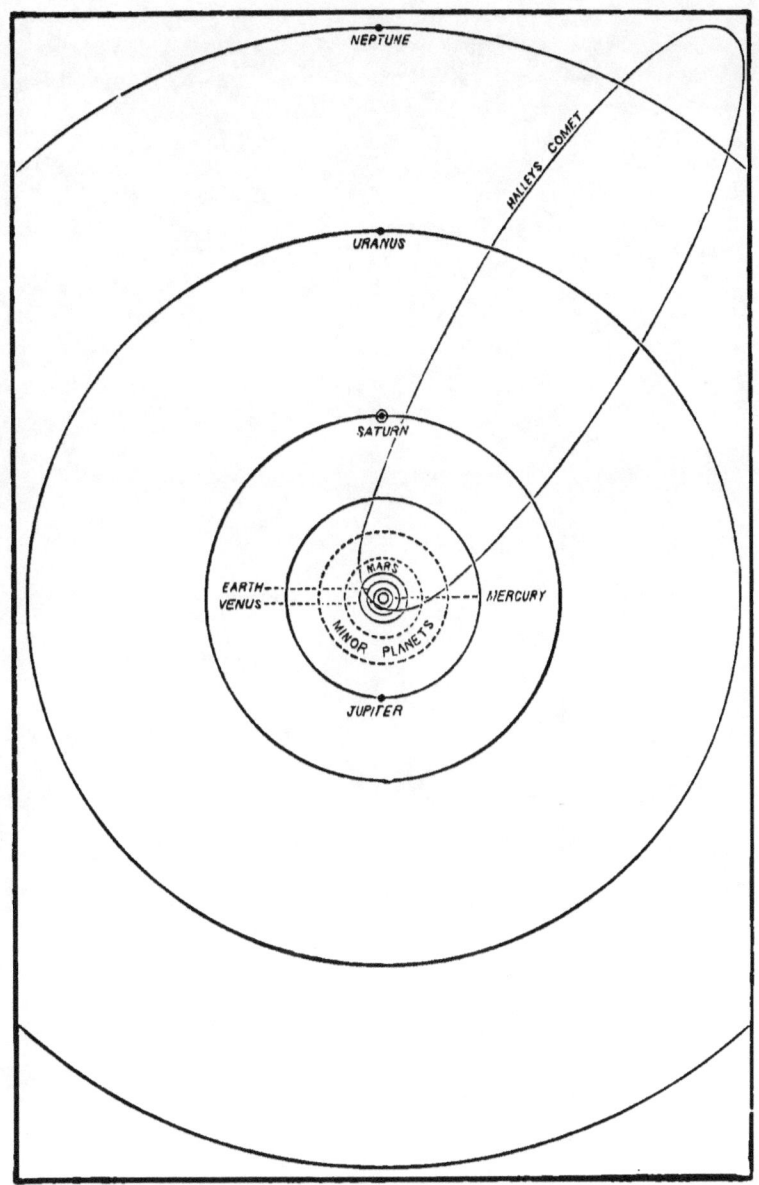

ORBITS OF THE PLANETS.

Outside the prominences lies the corona—a beautiful crown of soft light, plainly visible to the naked eye during an eclipse. Some of the streamers of this mysterious surrounding crown, or atmosphere, reach to a distance of a million miles. Little is yet known as to its true nature and use.

To return to the subject of sun-spots.

Although we cannot yet speak positively as to the constitution of the sun, we can at least say that he is a vast centre of light and heat, and a scene of terrific storms, beyond anything ever seen on this quiet little earth of ours. Certainly no such thing as quietness may be found there! The wildest and fiercest turmoil of fire and hurricane prevails unceasingly; and when the blazing surface is torn open, darker though still fiery depths seem to be exposed, becoming apparent as "spots" to us in our distant world. Beside the dark spots there are also spots of dazzling brilliance, standing out upon even that radiant background. Such extra-bright spots, which come and go, and at times change their form with great rapidity, are named *faculæ*—a Latin word meaning "torches."

It sounds to us startling to read of jets of fire and smoke from Mt. Vesuvius ten thousand feet in height, or of a river of lava in Iceland pouring in an unbroken stream for fifty miles. But what shall be thought of tongues of glowing hydrogen mounting to a height of one or two hundred thousand miles above the sun-surface? What shall be thought of a solar plane long enough to be folded three or four times around our solid earth? What shall be thought of the awful rush of incandescent gases, borne along at a rate of one or two or even three hundred miles in a second? What shall be thought of the vast rents in this raging ocean, rents often from fifty to one hundred thousand miles across?

Fifty thousand miles! A mere speck, seen from earth, scarcely visible without a telescope, yet large enough to contain seven worlds like ours flung in together. One of the bigger spots measured was so enormous that eighteen earths might have been laid in a row across the breadth of it; and to have filled up the entire hole about one hundred earths would have been necessary.

Only in speaking of a "hole" we must not picture a cavern, with solid walls and floor. If a sunspot be a hollow at all, which is not certain, it is a hollow framed and walled by glowing gases, in a state of inconceivable turmoil, heat, and fury.

CHAPTER III.

WHAT BINDS THE FAMILY TOGETHER?

" Dost thou know . . . the wondrous works of Him which is perfect in knowledge?" JOB 37:16.
"The day is Thine, the night also is Thine; Thou hast prepared the light and the sun. . . . Thou hast made summer and winter."—PSA. 74: 16, 17.

WHAT is it which binds together all the members of the Solar System?

Ah, what? Why should not the sun at any moment rush away in one direction, the earth in a second, the planets in half-a-dozen others? What is there to hinder such a catastrophe?

Nothing—except that they are all held together by a certain close family tie; or, more correctly, by the powerful influence of the head of the family.

This mysterious power which the sun has, and which all the planets have also in their smaller degrees, is called Attraction. Sometimes it is named Gravitation or Gravity.

When we speak, as we often do, of the *law* of attraction or gravitation, we mean simply this—

that throughout the universe, in things little and great, is found a certain wonderful *something* in constant action which we call a "law." What the "something" may be man cannot tell, for he knows it only by its effects. But these effects are seen everywhere, on all sides, in the earth and in the universe. It is well named in being called a "law," for we are compelled to obey it. None but the Divine Lawgiver who made this law could for a single moment interrupt its working.

What causes an apple to fall to the ground when it drops from the branch? Why should it not instead rise upwards? Because, of course, it is heavy, or has weight. But what *is* weight?

Simply this: that the earth draws or drags everything downwards towards herself by the power of attraction. Every substance, great or small, light or heavy, is made up of tiny atoms. Each one of these atoms attracts or draws all the other atoms towards itself; and the closer they are together, the more strongly they pull one another.

The atoms in a piece of iron are much closer than the atoms in a piece of wood; therefore the iron is called the "more *dense*" of the two, and its weight or "mass" is greater. The more closely the atoms are pressed together, the greater the number of them in a small space and the more strong the drawing towards the earth, for the earth draws each one of these atoms equally. That is only another way of saying that a thing is "heavier."

If you drop a stone from the top of a cliff will it rise upwards or float in the air? No, indeed. The pull of the earth's attraction, dragging and still dragging downward, makes it rush through the air, with speed quickening each instant, till it strikes the ground.

Every single atom in every single body *pulls* every other atom, whether far or near. The nearer it is, the stronger always the pulling.

We do not always feel this, because the very much greater attraction of the earth hides, or smothers as we may say, the lesser attraction of each small thing for another. But though you and I might stand side by side upon earth and feel no mutual attraction, yet if we could mount up a few thousands of miles, far away from earth, and float in distant space, there we should find ourselves drawn together and unable to remain apart.

Now precisely as an apple falling from a tree and a stone dropping from a cliff are dragged downward to the earth, just so our earth and all the planets are dragged downward towards the sun and towards each other. The law of the earth's attraction of all objects on its surface to itself was indistinctly suspected a very long time ago; but it was the great Newton who first discovered that this same law was to be found working among the members of the whole Solar System.

The sun attracts the earth, and the earth attracts the sun. But the enormous size of the sun compared with our earth—like a great nine-feet globe beside a tiny one-inch ball—makes our power of attraction to be quite lost sight of in his, which is so much greater.

We come now to another question. If the sun is pulling with such power at the earth and all her sister planets, why do they not fall down upon him? What is to prevent their rolling some day into one of those deep rents in his fiery envelope?

Did you ever tie a ball to a string and swing it rapidly round and round your head?

If you did, you must have noticed the steady

outward pull of the ball. The heavier the ball and the more rapid its whirl, the stronger the pull will be. Let the string slip, and the rush of the ball through the air to the side of the room will make this yet more plain.

Did you ever carry a glass of water quickly along, and then, on suddenly turning a corner, find that the water has not turned with you? It has gone on in its former direction, leaving the glass and spilling itself on the floor.

The cause in both cases is the same. Here is another "law of nature," so called. Though we can neither explain nor understand why and how it is so, we see it to be one of the fixed rules of nature, working everywhere alike throughout the whole universe.

The law, as we see it, seems to be this: Everything which is at rest must remain at rest until set moving by some cause outside of or independent of itself; and everything which is once set moving must continue moving in a straight line until checked.

According to this, a cannon-ball lying on the ground ought to remain there until it is set in motion; and, once set in motion by being fired from a cannon, it ought to go on for ever.

Exactly so — if nothing stops it. But the earth's attraction draws the cannon-ball downward, and every time it strikes the ground it is partly checked. Also each particle of air that touches it helps to bring it to rest. If there were no earth and no air in the question, the cannon-ball might rush on in space for thousands of years.

Why did the water get spilled?

Because it necessarily continued moving in a straight line. Your sudden change of direction compelled the solid glass to make the same change, but the liquid water was free to go straight on in its former course, so it obeyed this law and *did* go on.

Why did the ball pull hard at the string as you swung it round?

Because at each instant it was striving to obey this same law and to rush onward in a straight line. The pull of the string was every moment fighting against that inclination and forcing the ball to move in a circle.

Just such is the earth's movement in her yearly journey around the sun. The string holding in the ball pictures the sun's attraction holding in the earth. The pulling of the ball outward in order to continue its course in a straight line pictures the pulling of our earth each moment to break loose from the sun's attraction and to flee away into distant space.

For the earth is not at rest. Each tick of the clock she has sped onward over more than eighteen miles of her pathway through the sky. Every instant the sun is dragging, with all the great force of his attraction, to make her fall nearer to him. Every instant the earth is dragging, with all the great force of her rapid rush, to get away from him. These two pullings so far balance each other, or, more strictly, so far combine together, that between the two she journeys steadily round and round in her nearly circular orbit.

If the sun pulled a little harder she would need to travel a little faster, or she would gradually go nearer to him. If the earth went faster, and the sun's attraction remained the same as it is now, she would gradually widen her distance.

Indeed, it would only be needful for the earth to quicken her pace to about five-and-twenty miles a second, the sun's power to draw her being unchanged, and she would then wander away from him for ever. Day by day we on our earth should travel farther and farther away, leaving behind us all light, all heat, all life, and finding ourselves slowly lost in darkness, cold, and death.

For what should we do without the sun? All our light, all our warmth, come from him. Without the sun, life could not exist on the earth. Plants, herbage, trees would wither; the waters of rivers, lakes, oceans would turn to masses of ice; animals and men would die. Our earth would soon be one vast, cold, forsaken tomb of darkness and desolation.

I have spoken before about the old-world notion that our earth was a fixed plane, with the sun circling round her.

When the truth dawned slowly upon some great minds, anxious only to know what really was the truth, others made a hard struggle for the older and pleasanter mode of thinking. It went with many sorely against the grain to give up all idea of the earth being the chief place in the universe. Also there was something bewildering and dizzying in the notion that our solid world is never for one moment still.

But truth won the victory at last. Men consented slowly to give up the past dream and to

What Binds the Family Together? 49

learn the new lesson put before them. We still talk of the sun rising and setting and of the stars doing the same. This is, however, merely a common form of speech, which means just the opposite. For instead of the sun and stars moving, it is the earth which moves.

The earth has two distinct movements. Indeed, I ought to say that she has three, but we will leave all thought of the third for the present.

First: She turns round upon her axis once in every twenty-four hours.

Secondly: She travels round the sun once in about every three hundred and sixty-five days and a quarter.

No wonder our ancestors were startled to learn that the world, which they had counted so immovable, was perpetually spinning like a humming-top and rushing through space like an arrow.

You may gain some clear notions as to the daily rising and setting of our sun with the help of an orange. Pass a slender knitting-needle through the orange from end to end, and hold it about a yard distant from a single candle, in a room otherwise darkened. Let the needle or axis slant somewhat, and turn the orange slowly round and round upon it.

The candle does not move; but as the orange turns, the candle-light falls in succession upon each portion of the yellow rind. Half of the orange is always in shade and half is always in light; while at either side, if a small fly were standing there, he would be passing out of shade into candle-light or out of candle-light into shade.

Each spot on our earth moves round in turn into half-light, full-light, half-light, and darkness; or, in other words, has morning dawn, midday light, evening twilight, and night. Each spot on our earth would undergo regularly these changes every twenty-four hours throughout the year, were it not for another arrangement which so far affects this that the North and South Poles are, by turns, cut off from the light during many months together.

Thus the sun is in the centre of the Solar System, turning slowly on his axis; and the earth and the planets travel round him, each spinning like a teetotum, so as to make the most of his bright warm rays. But for this spinning movement of the earth, our day and night, instead of being each a few hours long, would each last six months.

You may notice that, as you turn the orange steadily round, the outside surface of the skin has to move much more slowly in those parts close to the knitting-needle than in those parts which bulge out farthest from it. Near the North and South Poles the surface of our earth travels slowly round a very small circle in the course of twenty-four hours. But at the equator every piece of ground has to travel about twenty-five thousand miles in the same time; so that it rushes along at the rate of more than one thousand miles an hour. A man standing on the earth, at the equator, is being carried along at this great speed, not *through* the air—for the whole atmosphere partakes of the same rapid motion—but *with* the air, round and round the earth's axis.

Now about the movement of the earth—her yearly journey round the sun.

While she moves, the sun, as seen from the earth, seems to change his place. First he is observed against a background of one group of stars, then against a second, then against a third. Not that the stars are visible in the day-time when the sun is shining, but their places are well known in the heavens; and also they can be noted very soon after he sets or before he rises, so that the constellations nearest to him may each day be easily found out.

Of course in old times the sun was thought to be really taking this journey among the stars, and men talked of "the sun's path" in the heavens. This path was named "the Ecliptic," and we use the word still, though we know well that the movements are not really his, but ours.

As the earth's daily movement causes day and night, so the earth's yearly movement causes spring, summer, autumn, and winter.

A few pages back I mentioned in passing one slight yet important fact which lies at the root of this matter about the seasons.

The earth, journeying round the sun, travels with her axis *slanting*.

Put your candle in the middle of the table, and stand at one end, holding your orange. Now let the knitting-needle, with the orange upon it, so slant that one end shall point straight over the candle, towards the upper part of the wall at the farther end of the room. Call the upper end, so pointing, the North Pole of your orange.

You will see that the candle-light falls chiefly upon the upper half of the orange; and as you turn it slowly, to picture day and night, you will find that the North Pole has no night and the South Pole has no day. That is summer in the northern hemisphere and winter in the southern.

Walk round next to one side of the table, towards the right hand, taking care to let the knitting-needle point steadily still in exactly the same direction, not towards the same *spot*, for that would alter its direction as you move, but towards the same *wall*. Stop, and you will find the candle lighting up one half of your orange, from the North to the South Poles. Turn it round slowly, never altering the slope of the axis, and you will see that every part of the orange comes by turns under the light. This is the Autumnal Equinox, when days and nights all over the world are equal in length.

Walk on to the other end of the table, still letting the needle slope and point steadily as before. Now the candle-light will shine upon the lower or South Pole, and the North Pole will be entirely in the shade. This is summer in the southern hemisphere and winter in the northern.

Pass on to the fourth side of the table, and

once more you will find it as at the second side—
equal light from North Pole to South Pole. This
will be the Spring Equinox.

It is an illustration that may be easily prac-
tised; but everything depends upon keeping the
slant of the needle or axis unchanged throughout.
If it be allowed to point first to right and then to
left, first towards the ceiling and then towards the
wall, the attempt will prove a failure.

NEPTUNE. SATURN. JUPITER.

CERES. MERCURY. MARS. VENUS. EARTH. URANUS.

COMPARATIVE MAGNITUDES OF THE PRIMARY PLANETS.

SUN, MOON AND STARS,

p. 64.

CHAPTER IV.

THE LEADING MEMBERS OF OUR FAMILY. — FIRST GROUP.

"The heavens declare the glory of God, and the firmament showeth His handywork."—PSA. 19:1.

THE chief distinction between stars and planets is, as before said, that the stars shine entirely by their own light, while the planets shine chiefly, if not entirely, by reflected light.

The stars are suns, great globes of heat and light. The planets simply receive the light of the sun, and shine with a brightness not their own.

A lamp shines by its own light; but a looking-glass, set in the sun's rays and flashing beams in all directions, shines by reflected light. In a dark room it would be dark. If there were no sun to shine upon Mars or Venus, we should see no brightness in them. The moon is like the planets in this. She has only borrowed light to give, and none of her own.

Any one of the planets, removed to the distance of the nearest fixed star, would be invisible to us. Reflected light will not shine nearly so far as the direct light of a burning body. There may be thousands or millions of planets circling round the stars—those great and distant suns—just as our brother-planets circle round our sun, but it is impossible for us to see them. The planets which we can see are close neighbors compared with the stars. I do not mean that they are near in the sense in which we speak of nearness upon earth. They are only near in comparison with what is so very much farther away.

For a while we must now leave alone all thought of the distant stars, and try to gain a clear idea of the chief members of our own family circle—that family circle of which the sun is the head, the centre, the source of life and warmth and light.

There are two ways in which astronomers group the planets of the Solar System.

One way is to divide them into the Inferior Planets and the Superior Planets.

As the earth travels in her pathway round the sun, two planets travel on their pathways round the sun nearer to him than ourselves. If the pathway or orbit of our earth were pictured by a hoop laid upon the table, with a ball in the centre

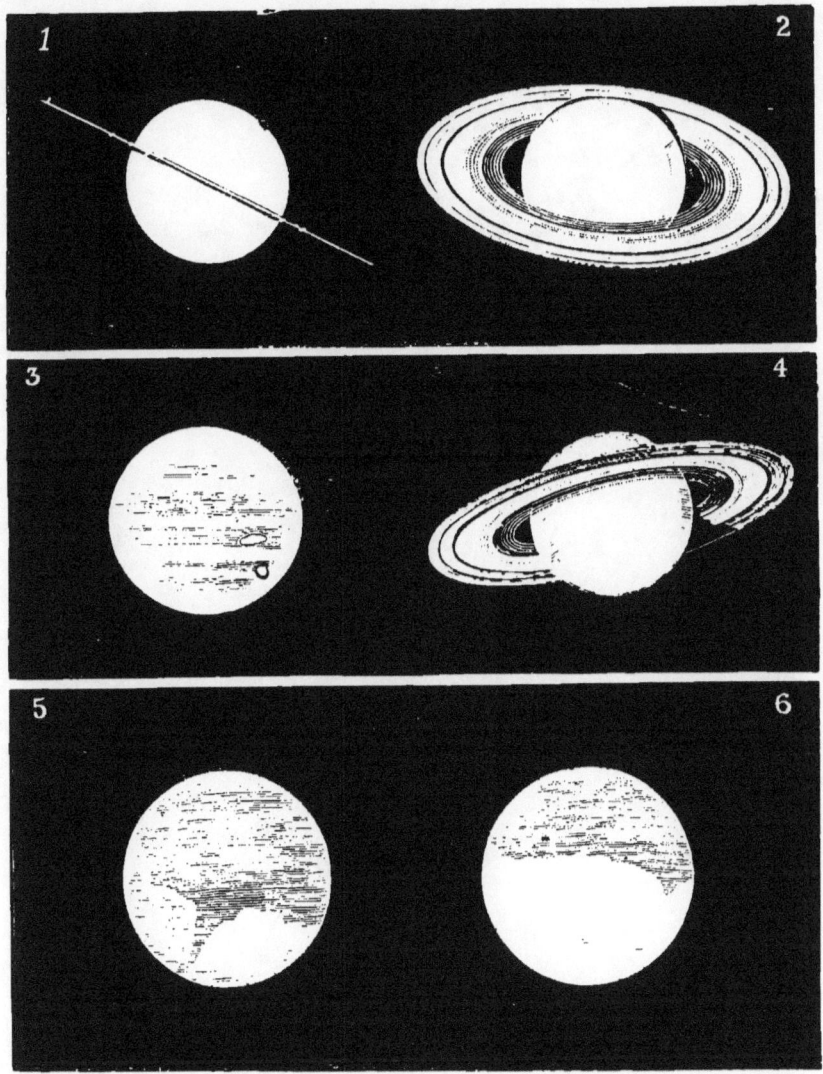

Figs. 1, 2, and 4, Saturn. In Fig. 1 the ring is represented as seen edgeways. Fig. 3. Jupiter. The dark spot represents a satellite passing across the disc of the planet.

Figs. 5 and 6, Mars; two views of nearly opposite hemispheres. The darker markings are believed to be water, and the lighter portion land. The white patch at the pole is probably ice or snow.

The Leading Members of our Family. 57

for the sun, then those two planets would have two smaller hoops of different sizes *within* ours, and the rest would have larger hoops of different sizes *outside* ours. The two within are called inferior planets, and the rest outside are called superior planets.

A round hoop would not make a good picture of an orbit. For the yearly pathway of our earth is not in shape perfectly round, but slightly oval; and the sun is not exactly in the centre, but a little to one side of the centre. This is more or less the case with the orbits of all the planets.

But the laying of the hoops upon the table would give no bad idea of the way in which the orbits really lie in the heavens. The orbits of all the chief planets do not slope and slant round the sun in all manner of directions. They are placed almost in the same *plane*, as it is called—or, as we might say, in the same *flat*. In these orbits the planets all travel round in the same direction. One may overtake a second on a neighboring orbit, and get ahead of him, but one planet never goes back to meet another.

In speaking of the orbits, I do not mean that the planets have visible marked pathways through the heavens, any more than a swallow has a visible pathway through the sky, or a ship a marked pathway through the sea. Yet each planet has his own orbit, and each planet so distinctly keeps to his own that astronomers can tell us precisely whereabouts in the heavens any particular planet will be, at any particular time, long years beforehand.

There is also another mode of grouping the planets besides dividing them into superior and inferior planets.

By this other mode we find two principal groups or quartettes of planets, separated by a zone or belt of a great many very small planets.

First Group ....
- Mercury.
- Venus.
- Earth.
- Mars.

The Asteroids or Planetoids.

Second Group ..
- Jupiter.
- Saturn.
- Uranus.
- Neptune.

The first four are small compared with the last four, though much larger than any in the belt of tiny Asteroids.

It was believed at one time that a planet had

been discovered nearer to the sun than Mercury, and the name Vulcan was given to it. But no more has been seen of Vulcan, and his existence is so doubtful that we must not count him as a member of the family without further information.

Mercury is very much smaller than our earth.

The diameter of the earth is eight thousand miles, but the diameter of Mercury is less than three thousand miles—not even half that of the earth.

Being so much nearer to the sun than ourselves, the pulling of his attraction is much greater, and this has to be balanced by greater speed, or Mercury would soon fall down upon the sun. Our distance from the sun is ninety-two millions of miles. Mercury's distance is only about one-third of ours; and instead of travelling, like the earth, at the rate of eighteen miles each second, Mercury, when nearest to the sun, goes at the mad pace of thirty-five miles each second. It is a good thing the earth does not follow his example, or she would soon break loose from the sun's control altogether.

The earth takes more than three hundred and

sixty-five days, or twelve months, to journey round the sun in her orbit. That is what we call "the length of our year." But Mercury's year is hardly eighty-eight days, or not quite three of our months. No wonder!—when his pathway is so much shorter and his speed so much greater than ours. So Mercury has four years to one year on earth; and a person who had lived on Mercury as long as five earthly years would then be twenty years old. The increased number of birthdays would scarcely be welcome in large families, supposing we could pay a long visit there.

The sun, as seen from Mercury, looks about four and a half times as large as from here, the heat and glare being increased in proportion. No moon has ever been found belonging to Mercury.

Venus, the second inferior planet, is nearly the same size as our earth. Seen from the earth, she is one of the most brilliant and beautiful of all the planets. Her speed is four miles a second faster than ours, and her distance from the sun is about two-thirds that of our own; so that the orbit of Venus lies half-way between the orbit of Mercury and the orbit of the earth. Her year is nearly two hundred and twenty-five days, or seven and a half of our months. Some astronomers at one time thought that they caught glimpses of a moon near Venus: but this is still quite doubtful; and indeed it is believed to have been a mistake, since for a long while it has not been seen again.

Venus and Mercury are only visible as morning and evening planets. Venus, being farther from the sun, does not go before and follow after him quite so closely as Mercury, and she is therefore the longer within sight.

When Venus, travelling on her orbit, comes just between the sun and us, her dark side is turned towards the earth, and we can catch no glimpse of her. When she reaches that part of her orbit which is farthest from us, quite on the other side of the sun, her great distance from us makes her light seem less. But about half-way round on either side she shows exceeding brilliancy; and that is the best view we can get of her.

Seen through a telescope, Venus undergoes what we call "phases," like the moon. That is to say, we really have "new Venus," "quarter Venus," "half Venus," "full Venus," and so on. Mercury passes through the same phases, but his smallness and distance make them more difficult to see.

Next to the orbit of Venus comes the orbit of our own Earth, the third planet of the first group.

Mars, the fourth of the inner quartette, but the first of the superior planets, is a good deal smaller than Venus or the earth. The name Mars, from the heathen god of war, was given on account of his fiery reddish color. Mars is better placed than Venus for being observed from earth. When he is at the nearest point of his orbit to us we see him full in the blaze of sunlight; whereas Venus, at her nearest point, turns her bright face away.

The length of the day of Mars, or in other words the time he takes to turn upon his axis, is only forty minutes longer than that of the earth.

Mars' journey round the sun is completed in the course of six hundred and eighty-seven days, not much less than two of our years. His distance from the sun is about one hundred and forty millions of miles, and his speed is fourteen miles a second. We shall find, with the increasing distance of each planet, that the slower pace balances the lessened amount of the sun's attraction.

Passing on from Mars, the last of the first group of planets, we reach the belt of Asteroids, sometimes called Planetoids, Minor Planets, or Telescopic Planets. They are so tiny that Mercury is a giant compared with the largest among them.

The zone of space containing all these little planets is more than a hundred millions of miles broad. Their orbits do not lie flat in almost the same plane, but slant about variously in a very entanged fashion. If a neat model were made of this zone, with a slender piece of wire to represent each orbit, it would be found impossible to lift up one wire without pulling up all the rest with it. Those asteroids lying nearest to the sun take about three of our years to travel round him, and those lying farthest take about six of our years.

New members of the group are very often

found. The number of asteroids now known amounts to nearly three hundred and fifty.

Pallas, the largest, is about six hundred miles in diameter. Vesta, the brightest, is about three hundred miles. Nearly eighteen thousand Vestas would be needed to make one globe equal to our earth in size.

CHAPTER V.

THE LEADING MEMBERS OF OUR FAMILY.— SECOND GROUP.

"O Lord, how manifold are Thy works! in wisdom hast Thou made them all." Psa. 104:24.

LEAVING behind us the busy zone of Planetoids hurrying round and round the sun in company, we cross a wide gap and come upon a very different sight.

The distance from the sun which we have now reached is more than four hundred and eighty millions of miles, or over five times as much as the earth's distance; and the sun in the heavens shows a diameter not one-fifth of that which we are accustomed to see. Slowly — yet not slowly—floating onwards through space in his far-off orbit, we find the magnificent planet Jupiter.

Is eight miles each second slow progress? Compared with the wild whirl of little Mercury, or even compared with the rate of our own earth's advance, we may count it so; but certainly not compared with our notions of speed upon earth.

Eight miles each second is five hundred times as fast as the swiftest express train ever made by man. No mean pace that for so enormous a body!

For Jupiter is the very largest of all the members of the Solar System except the sun himself—quite the eldest brother of the family. His longest diameter is nearly eighty-eight thousand miles, or about eleven times as long as that of earth. Though, in proportion to his great bulk, not nearly so heavy as our earth, yet his bulk is so vast that more than twelve hundred earths would be needed to make one Jupiter.

It must not, however, be forgotten that there is a certain amount of uncertainty about these measurements of Jupiter. He seems to be so covered with a dense atmosphere and heavy clouds that it is quite impossible for us to learn the exact size of the solid body within—if, indeed, any part of him is solid.

Jupiter does not travel alone. Borne onwards with him and circling round him are four or five moons; one about the same size as our own moon, and of the others three are larger. The nearest of the four, which till lately were counted to be the full number, speeds round him in less than two of our days. The most distant, though over a million miles away, takes scarcely seventeen days to accomplish its long journey. A fifth moon is now believed to exist.

Jupiter and his moons make quite a little system by themselves—a family circle within a family circle.

Like the smaller planets, Jupiter spins upon his axis; and he does this so rapidly that, notwithstanding his great size, his day lasts only ten hours instead of twenty-four hours like ours.

But if Jupiter's day is short, his year is not. Nearly twelve of our years pass by before Jupiter has travelled once completely round the sun. So a native of earth who had just reached his thirty-seventh year would, on Jupiter, be only three years old.

Passing onward from Jupiter, ever farther and farther from the sun, we leave behind us another vast and empty space—empty as we count emptiness, though it may be that there is in reality no such thing as emptiness throughout the length and breadth of the universe.

The width of the gap which divides the pathway of Jupiter from the pathway of his giant brother-planet Saturn is nearly five times as much as the width of the gap separating the earth from the sun. The distance of Saturn from the sun is not much less than double the distance of Jupiter.

With this great space in our rear we come upon another large and radiant planet, the centre, like Jupiter, of another little system, though it can only be called "little" in comparison with the much greater Solar System of which it forms a part.

Saturn's diameter is less than that of Jupiter, but the two come near enough to be naturally ranked together. Nearly seven hundred earths would be needed to make one globe as large as Saturn. But here again the dense and cloudy envelope makes us very uncertain about the planet's actual size. Saturn is like Jupiter in being made of lighter materials than our earth; and also in his rapid whirl upon his axis, the length of his "day" being a little over ten of our hours.

From Jupiter's speed of eight miles each second we come down in the case of Saturn to about five miles each second. And Jupiter's long annual journey looks almost short seen beside Saturn's longer journey of almost thirty years. A man aged sixty, according to our fashion of reckoning time, would on Saturn have just kept his second birthday.

The system or family of Saturn is yet more wonderful than that of Jupiter. Not five only, but eight, moons travel ceaselessly around Saturn, each in its own orbit; and in addition to the eight moons he has revolving round him three magnificent rings. These rings, as well as the moons, shine, not by their own brilliancy, for they have none, but by borrowed sunlight.

The farthest of the moons wanders in his lonely pathway about two millions of miles away from Saturn. The largest of them is believed to be about the same size as the planet Mars.

Of the three rings circling round Saturn, almost exactly over his equator, the inside one is dusky, purplish, and transparent; the one outside or over that is very brilliant; and the third, outside the second, is rather grayish in hue.

Another vast gap—more enormous than the last. It is a wearisome journey. From the orbit of Jupiter to the orbit of Saturn at their nearest points was five times as much as from the sun to the earth. But from the orbit of Saturn to the orbit of Uranus, the next member of the sun's family, we have double even that great space to cross.

Still, obedient to the pulling of the sun's attractive power, Uranus wanders onward in his wide pathway round the sun at the rate of four miles per second. Eighty-four of our years make one year of Uranus. He is attended by four moons, and thus forms a third smaller system within the Solar System; but he may have other satellites also, as yet undiscovered. In size he is sixty-four times as large as our earth.

One more mighty chasm of nine hundred millions of miles, for the same distance which separates the pathway of Saturn from the pathway of Uranus separates also the pathway of Uranus from the pathway of Neptune. Cold and dark and dreary indeed seems to us the orbit on which this banished member of our family circle creeps round the sun, in the course of one hundred and sixty-five years, at the sluggish rate of three miles per second.

On the planet Saturn the quantity of light and heat received from the sun is not much more than

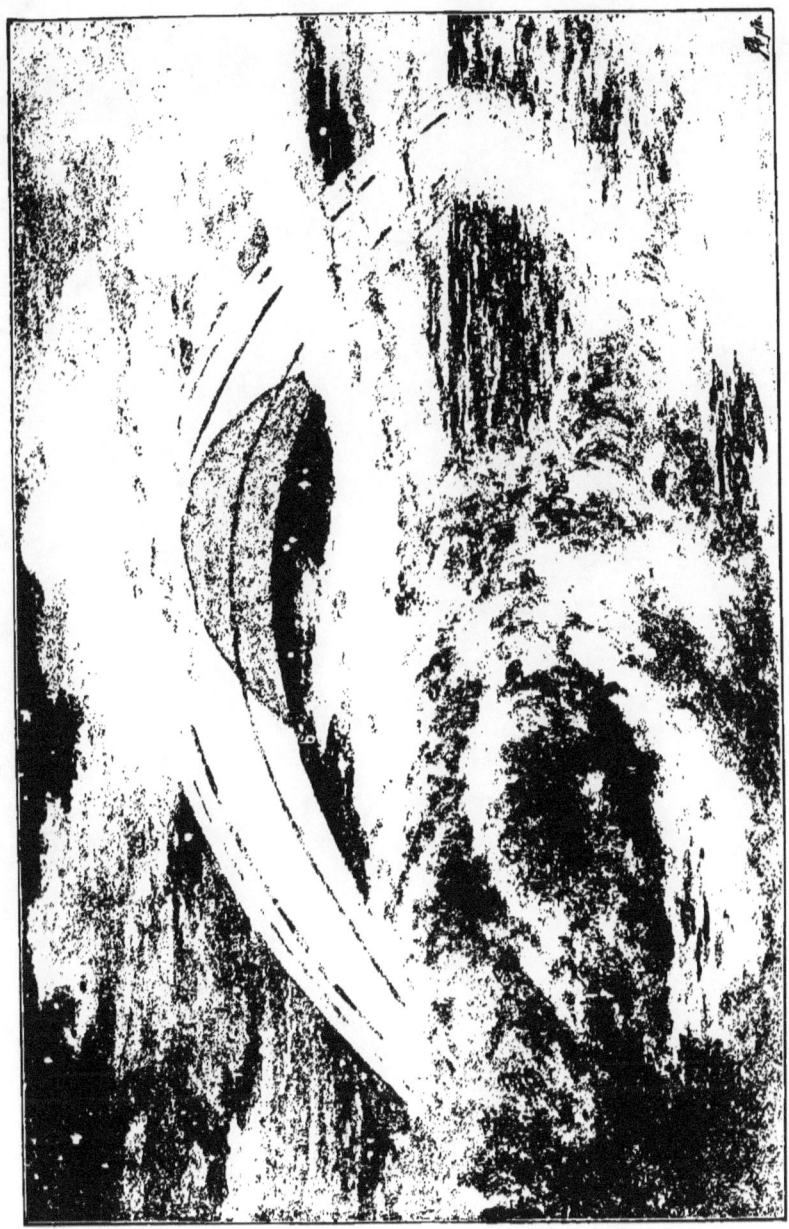

a hundredth part of that which we are accustomed to receive on earth. But by the time we reach Neptune the great sun has faded and shrunk in the distance until to our eyes he looks only like an exceedingly brilliant and dazzling star.

We know little of this far-off brother, Neptune, except that he is rather larger than Uranus, being nearly thirty-five thousand miles in diameter; that he has at least one moon; and also that, like Uranus, he is made of materials lighter than those of earth, but heavier than those of Jupiter or Saturn.

After all, it is no easy matter to gain clear ideas as to sizes and distances from mere statements of "so many miles in diameter" and "so many millions of miles away." A "million miles" carries to the mind a very dim notion of the actual reality.

Now if we can in imagination bring down all the members of the Solar System to a small size, keeping always the same proportions, we may find it a help.

"Keeping the same proportions" means that all must be lessened alike, all must be altered in the same degree. Whatever the supposed size of the earth may be, Venus must be still about the same size as the earth, Saturn seven hundred times as large, and so on. Also, whatever the distance of the earth from the sun, in miles or yards or inches, Mercury must still be one-third as far, Jupiter still five times as far, and thus with the rest.

First as to size alone. Suppose the earth is brought down to a small globe exactly three inches in diameter. It will be a very smooth globe. Not only men and houses, but mountains, valleys, seas, will all have shrunk to so minute a size as to be quite invisible to the naked eye.

Fairly to picture the other members of the Solar System in due proportion, you will have them as follows:*

Mercury and Mars will be balls smaller than the earth, and Venus nearly the same size as the earth. Uranus and Neptune will be each somewhere about a foot in diameter. Saturn will be twenty-eight inches, and Jupiter thirty-two inches, in diameter. The sun will be a huge dazzling globe *twenty-six feet* in diameter. No wonder he weighs seven hundred and fifty times as much as all his planets put together!

* " Other Worlds," by R. A. Proctor.

Next let us picture the system more exactly on another and smaller scale.

First think of the sun as a brilliant globe two feet in diameter, floating in space far from this earth of ours.

At a distance of about eighty-two feet travels a tiny ball, no bigger than a grain of mustard-seed, passing slowly round the sun—*slowly*, because, as size and distance are lessened, speed must in due proportion be lessened also. This is Mercury.

At a distance of about one hundred and forty-two feet from the central sun travels another tiny ball, the size of a pea. This is Venus.

At a distance of two hundred and fifteen feet from the sun travels a third tiny ball, again the size of a pea. This is the Earth; and her accompanying Moon is no larger than a very small pin's head.

At a distance of three hundred and twenty-seven feet from the sun travels a fourth tiny ball, the size of a rather large pin's head. This is Mars.

Then comes a wide blank space, followed by a large number of minute grains of sand floating round the sun in company, at distances varying from five hundred to six hundred feet. These are the Asteroids. Another wide blank space succeeds the outermost of them.

Nearly a quarter of a mile distant from the sun journeys a globe very different from those last described, being as large as a moderate-sized orange. Round him and with him, as he rolls slowly onward, travel five smaller balls, the biggest of them not much over a mustard-seed in size. These are Jupiter and his moons.

Somewhat less than half a mile distant from the sun journeys another globe, about the size of a small orange. Eight tiny balls and three delicate rings circle slowly round him as he moves. These are Saturn and his belongings.

More than three-quarters of a mile distant from the sun journeys another globe, about the size of a small plum, accompanied by four minute balls. These are Uranus and his moons.

Lastly, at a distance of about a mile and a quarter from the sun, journeys one more globe, as large as a good-sized plum, with one tiny companion, pursuing his far-off journey.

These proportions as to size and distance,* if carefully studied, will serve to give a clearer idea of the Solar System as a whole than learning a long list of numbers would do.

* From " Outlines of Astronomy," by Sir J. Herschel.

CHAPTER VI.

OUR PARTICULAR FRIEND AND ATTENDANT.

"The moon walking in brightness."—JOB 31 : 26.

COME and let us pay a visit to the moon.

We seem to feel a personal interest in her just because she is in so peculiar a sense our own friend and close attendant. The sun shines for us, but then he shines for all the members of the Solar System. And the stars—so many as we can see of them—shine for us too, but no doubt they shine far more brilliantly for other and nearer worlds. The moon alone seems to belong especially to ourselves.

Indeed we are quite in the habit of speaking about her as "our moon." Rather a cold and calm friend some may think her, sailing always serenely past, whatever may be going on beneath her beams; yet she has certainly proved herself constant and faithful in her attachment.

We have not very far to travel before reaching her—merely about two hundred and forty thousand miles. That is nothing compared with the

Our Particular Friend and Attendant. 77

weary millions of miles which we have had to cross to visit some members of our family. A rope two hundred and forty thousand miles long would fold neatly ten times round the earth at the equator. You know the earth's diameter—about eight thousand miles. If you had thirty poles, each eight thousand miles long, and could fasten them all together, end to end, one beyond another, you would have a rod long enough to reach from the earth to the moon.

Let us take a good look at her before starting.

She is very beautiful. That soft silvery light, so unlike sunlight or gaslight or any other kind of light, even electric light, has made her the darling of poets and the delight of all who love nature. Little children like to watch her curious markings, and to make out the old man with his bundle of sticks, or the eyes, nose, and mouth of the moon—not dreaming what those markings really are. And in moods of sadness how the pure calm moonlight seems to soothe the feelings! Who would suppose that the moon's beauty is the beauty rather of death than of life?

The stars have not much chance of shining

through her bright rays. It is well for astronomers that she is not always at the full.

But when she is how large she looks—quite as large as the sun, though in reality her size, compared with his, is only as a very small pin's head compared with a school globe two feet in diameter. Her diameter is little more than two thousand miles, or one-quarter that of our earth; and her whole surface, spread out flat, would scarcely equal North and South America without any of the surrounding islands.

The reason she looks the same size as the sun is that she is so very much nearer. The sun's distance from us is more than one-third as many *millions* of miles as the moon's distance is *thousands* of miles. This makes an enormous difference.

We call our friend a "moon," and say that she journeys round the earth, while the earth journeys round the sun. This is true, but it is only part of the truth. Just as certainly as the earth travels round the sun, so the moon also travels round the sun. And just as surely as the earth is a planet, so the moon also is a planet. It is a common mode of expression to talk about "the earth and her satellite." A no less correct, if not more correct, way would be to talk of ourselves as "a pair of planets," journeying round the same sun, each pulled strongly towards him, and each pulling the other with a greater or less attraction, according to her size and weight.

For the sun actually does draw the moon with more force than that with which the earth draws her. Only as he draws the earth with the same sort of force, and nearly in the same degree, he does not pull them apart.

The moon, like the other planets, turns upon her axis. She does this very slowly, however; and rather singularly she takes exactly the same time to turn once upon her axis that she does to travel once round the earth.

The result of this is that we only see one face of the moon. If she turned upon her axis and journeyed round the earth in two different lengths of time, or if she journeyed round us and did not turn upon her axis at all, we should have views of her on all sides, as of other planets. But as her two movements so curiously agree, it happens that we always have one side of the moon towards us, and never catch a glimpse of the other side.

And now we are ready to start on our journey of two hundred and forty thousand miles.

An express train, moving ceaselessly onward night and day at the rate of sixty miles an hour, would take us there in about five months and a half. But no line of rails has ever yet been laid from the earth to the moon, and no "lightning express" has ever yet plied its way to and fro on that path through the heavens. Not on the wings of steam, but on the wings of imagination, we must rise aloft. Come—it will not take us long. We shall pass no planets or stars on the road, for the moon lies nearer to us than any other of the larger heavenly bodies.

Far, far behind us lies the earth, and beneath our feet as we descend stretch the broad tracks of moonland. For "downward" now means towards the moon and away from the distant earth.

What a strange place we have reached! The weird ghastly stillness of all around, and the awful dazzling cloudless glare, strike us first and most forcibly. Nothing quite like this have we ever felt on the earth. The close of the moon's long day— on this side of its globe—is approaching, and during a whole fortnight past the sun's fierce rays have been beating down on these shelterless plains; yet, from lack of an atmosphere to retain sun-heat, the ground is cold—actually below zero—while the rays strike us with scorching power. It is much as on some lofty mountain-top, where icy cold and burning glare are combined; only here all is immoderately intensified.

Not a cloud to be seen overhead; only a sky of inky blackness, with a blazing sun and thousands of brilliant stars and the dark body of our own earth, large and motionless and rimmed with light. Seen from earth, sun and moon look much the same size; but seen from the moon the earth looks thirteen times as large as our full moon.

Not even a little mistiness in the air to soften this fearful glare! Air! why, there is no air; at least, not enough for any human being to breathe or feel. If there were air the sky would be blue, not black, and the stars would be invisible in the day-time. It looks strange to see them now shining beside the sun.

And then this deadly stillness! Not a sound, not a voice, not a murmur of breeze or water. How could there be? Sound cannot be carried without air, and of air there is none. As for breeze—wind is moving air, and where we have no air we can have no wind. As for water—if there ever was any water on the moon it has entirely disappeared. We shall walk to and fro vainly in search of it now. No rivers, no rills, no torrents in those stern mountain-ramparts rising on every side. All is craggy, motionless, desolate.

How very, very slowly the sun creeps over the black sky; and no marvel, since a fortnight of earth-time is here but one day, answering to twelve hours upon earth. Cannot we find shelter somewhere from this blaze of heat? Yonder tall rock will do, casting a sharp shadow of intense blackness. We never saw such shadows upon earth. There the atmosphere so breaks and bends and scatters about the light that outlines of shadows are soft and hazy, even the clearest and darkest of them, compared with this.

How soon will the sun go down? One could wish for lessened glare, though by no means for greater cold. Meanwhile he is well worth studying—through this piece of neutral-tinted glass, without which no human eyes could face his splendor. For here no atmosphere lies between to act as a sheltering veil. What a magnificent object he is, with his radiant photosphere and crimson border, out of which spring sharp-toothed prominences, visible without the help of a telescope.

No air between to hide them now, therefore no need of an eclipse to render them visible. See too the exquisite corona, a crown of pearly light, stretching far on all sides in delicate lines and streams which die out slowly against the jet background. The black spots on the face of the sun are very distinct, and so also are the brilliant faculæ.

We must take a look around us now at moonland, and not only sit gazing at the sun, though such a sky may well enchain attention.

How unlike our earthly landscapes! No sea, no rivers, no lakes, no streams, no brooks, no trees, bushes, plants, grass, or flowers, no wind or breeze; no cloud or mist or thought of possible rain; no sound of bird or insect, of rustling leaves or trickling water. Nothing but a changeless glare contrasting with inky shadows—sun and earth and stars in a black heaven above, silent desolate mountains and plains below.

For though we stand here upon a rough plain, this moon is a mountainous world. Ranges of rugged hills stretch away in the distance, with valleys lying between—not soft green sloping earthly valleys, but steep gorges and precipitous hollows, all white dazzle and deep shade.

But the mountains do not commonly lie in long

ranges as on earth. The surface of the moon seems to be dented with strange round pits, or craters, of every imaginable size. We had a bird's-eye view of them as we descended at the end of our long journey moonward. In many parts the ground appears to be quite honeycombed with them. Here are small ones near at hand and larger ones in the distance. The smaller craters are surrounded by steep ramparts of rock, the larger ones by circular mountain-ranges. We have nothing quite like them on earth.

Are they volcanoes? So it would seem, only no life, no fire, no action, remain now. All is dead, motionless, still. Is this verily a blasted world? Has it fallen under the breath of divine wrath, coming out scorched and seared? Or is it rather passing through an aged and used-up phase of existence, through which other planets also pass, or will pass, at some stage of their career? The latter seems probable.

We will move onward, and look more closely at that towering mass of rugged rocks beyond which the sun will by-and-by go down. Long jetty shadows lie from them in this direction. No wonder astronomers on earth can through their telescopes plainly see these black shadows contrasting with the glaring brightness on the other side.

A "mass of rocks," I have said; but as with our powers of rapid movement we draw near, we find a range of craggy mountains sweeping round in a vast circle. Such a height in Switzerland would demand many hours of hard climbing. But on this small globe attraction is a very different matter from what it is on earth. Our weight is so lessened that we can leap the height of a tall house without the smallest difficulty. No chamois ever sprang from peak to peak in his native Switzerland with such amazing lightness as that with which we now ascend these mighty rocks.

Ha! what a depth on the other side! We stand looking down into one of the monster craters of the moon. A sheer descent of about sixteen thousand feet would land us at the bottom. Why, Mont Blanc itself is only some fifteen thousand feet in height. And what a crater! Sixty-five miles across in a straight line from here to the other side, with these lofty rugged battlements circling round, while from the centre of the rough plain below a sharp cone-shaped mountain rises to about a quarter of the height of the surrounding range.

It is a grand sight; peak piled upon peak, crag upon crag, sharp rifts or valleys breaking here and there the line of the narrow uplifted ledge, all wrapped in silent and desolate calm. There are many such craters as this on the moon, and some much larger.

The sun slowly nears his setting and sinks behind the opposite range. How we shiver! The last ray of sunlight has gone, and the ground, already below zero, is fast growing colder, pouring out into space all its little remnant of warmth. The change takes place with marvellous quickness. A deadly chill creeps over all around. A whole fortnight of earth-time must pass before the sun's rays will again touch this spot. Verily the contrasts of climate in the moon, during the twelve long days and nights which make up her year, are, to say the least, unpleasant!

But though the sun is gone we are not in darkness. The stars shine with dazzling brightness, and the huge body of the earth, always seeming to hang motionless at one fixed point in the sky, gives brilliant light, though at present only half her face is lit up and half is in shadow. Still her shape is plainly to be seen, for she has ever round her a ring of light caused by the gathered shining of stars as they pass behind her thick atmosphere. She covers a space on the sky more than a dozen times as large as that covered by the full moon in our sky.

It would be worth while to stay here and watch the half-earth grow into magnificent full-earth. But the cold is becoming fearful—too intense for even the imagination to endure longer. What must be the state of things on the other side of the moon, where there is no bright earth-light to take the place of the sun's shining, during the long two-weeks' night of awful chill and darkness?

Time for us to wend our way homeward from this desolate hundredfold-arctic scene. We have more to learn by-and-by about our friend and companion. For the present—enough.

CHAPTER VII.

VISITORS.

"Thou knowest not the works of God who maketh all."
ECCLES. 11:5.

WE come next to the very largest members of our Solar System.

From time to time in past days—and days not very long past either—people were startled by the sight of a long-tailed star moving quickly across the sky, called a comet. We see such long-tailed stars still, now and then, but their appearance no longer startles us.

It is hardly surprising, however, that fears were once felt. The great size and brilliancy of some of these comets naturally caused large ideas to be held as to their weight; and the general uncertainty about their movements naturally added to the mysterious notions afloat with respect to their power of doing harm.

A collision between the earth and a comet seemed no unlikely event; and if it happened—what then? Why, then of course the earth would be overpowered, crushed, burned up, destroyed.

So convinced were many on this point that the sight of a comet and the dread of the coming "end of the world" were fast bound together in their minds.

Even when astronomers began to understand the paths of some of the comets, and to foretell their return at certain dates, the old fear was not quickly laid to rest. So late as the beginning of the present century, astronomers having told of an approaching comet, other people added the tidings of an approaching collision. "If a collision, then the end of the world," was the cry; and one worthy family, living and keeping a shop in a well-known town on the south coast of England, packed up and fled to America—doubtless under full belief that the destruction of the Old World would not include the destruction of the New.

The nature of these singular bodies is somewhat better known in the present day; yet even now among all the members of the Solar System they are perhaps the ones about which we have most to learn.

The nucleus or bright and star-like spot which, with the surrounding coma or "hair," we sometimes call the "head" of the comet, is the densest and heaviest part of the whole. Comets may be of enormous size, sometimes actually filling more space than the sun himself; and their tails stream often for millions of miles behind or before them; nevertheless they are among the lightest of the members of the Solar System.

This excessive lightness greatly lessens the comet's power of harm-doing. In the rebound from all the old exaggerated fears men laughed at the notion of so light and delicate a substance working any injury whatever, and even declared that a collision might take place without people on earth being aware of the fact. It is now felt that we really know too little about the nature of the said substance to be able to say what might or might not be the result of a collision. A slight amount of injury to the surface of the earth *might* possibly take place. But of the "end of the world" as likely to be brought about by any comet in existence we may safely banish all idea.

The word "comet" means "a hairy body," the name having been given from the hairy appearance of the light around the nucleus.

A great many hundreds of different comets have been seen at different times by men—some large, some small, some visible to the naked eye, but most of them only visible through telescopes.

These hundreds are, there is no doubt, but a very small number out of the myriads ranging through the heavens.

If you were seated in a little boat in mid-ocean, counting the number of fishes which in one hour passed near enough in the clear water for your sight to reach them, you might fairly conclude, even if you did not know the fact, that for every single fish which you could see there were tens of thousands which you could not see.

Reasoning thus about the comets, as we watch them from our earth-boat in the ocean of space, we feel little doubt that for each one which we can see millions pass to and fro beyond reach of our vision. Indeed, so long ago as the days of Kepler that great astronomer gave it as his belief that the comets in the Solar System, large and small, were as plentiful as the fishes in the sea. And all that modern astronomers can discover only tends to strengthen this view.

Why should the comets be called "visitors"?

I call them so simply because many of them *are* visitors. Some, it is true, belong to the Solar System. But even in their case strong doubts are felt whether they were not once visitors from a distance, caught in the first instance by the attraction of one of the larger planets and retained thenceforward, for a time at least, by the strong attraction of the sun.

Every comet, like every planet, has his own orbit or pathway in the heavens, though the kind of orbit varies with different comets.

There are, first, those comets which travel round and round the sun in "closed orbits"—that is, in a ring with joined ends—only the ring is always oval, not round.

There are, secondly, those which travel in an orbit which *may* be closed; but if so, the oval is so long and narrow, and the farther closed end is at so great a distance, that we cannot speak certainly.

There are, thirdly, those which are decidedly mere visitors. They come from the far-off star-depths, flash once through our busy Solar System, forming radiant trains of light as they draw nearer to the sun; and then they pass away in another direction, never to return.

Only a small part of the orbits of these comets can be seen from earth; but by careful attention astronomers learn something of the shape of the curve in which they travel. It is in that way possible to calculate, sometimes certainly and sometimes uncertainly, whether a comet may be expected to return, or whether we have seen him for the first and the last time. By looking at *part* of a curve, the rest of which is hidden from us, we are able to judge whether that part belongs to a circle or an oval, or whether the two ends pass away in different directions and do not join.

The comets, whether members of our family circle or visitors from a distance, are altogether very perplexing. They are often extremely large, yet they are always extremely light. They reflect the sun's brightness like a planet, yet in some measure they seem to shine by their own light, like a star. They obey the attraction of the sun, yet he appears to have a singular power of driving the comets' tails away from himself.

For, however rapidly the comet may be rushing round the sun, and however long the tail may be, it is almost always found to stream in an opposite direction from the sun. An exception to this rule was seen in the case of a certain comet with two tails, one of which did actually point towards the sun; but the inner tail may have been only a "jet" of unusual length, like in kind to the smaller jets often thus poured out from the nucleus.

Very curious changes take place in comets as they journey, especially as they come near the sun. One was seen in the course of a few days to lose both coma and tail, but as a general rule the tails increase with proximity to the sun, and lessen or vanish with distance. Travelling, as the comets do, from intense cold into burning heat, they are very much affected by the violent change of climate.

For the paths of the comets are such long ovals or ellipses that, while they approach the sun very closely in one part of their "year," they journey to enormous distances in the other part.

"Halley's Comet," which takes seventy-six of our years to travel round the sun, comes nearer to him than Venus and goes farther away from him than Neptune.

As this comet draws gradually closer he has to make up for the added pull of the sun's increasing attraction by rushing onward with greater and greater rapidity, till he whirls madly past the sun, and then, with slowly slackening speed, journeys farther and farther away, creeps at length lazily round the farther end of his orbit in the chill dark neighborhood of Neptune, and once more travels towards the sun with growing haste.

"Encke's Comet" has a year of only three and a half of our years, so he may be said to live quite in our midst. But many comets go very much farther away than the one named after Halley. It is calculated of some that, if they ever return at all, it cannot be for many hundreds of years.

"Newton's Comet," seen about two centuries ago, has a journey to perform of such length that he is not expected again to appear for several thousands of years. Yet at the nearest point in his orbit he approached the sun so closely that the heat which he endured was about two thousand times that of red-hot iron. Changes were seen to be taking place in his shape as he drew near to the sun and disappeared. Four days he was hidden in the sun's rays. He vanished with a tail streaming millions of miles behind him. He made his appearance again with a tail streaming millions of miles in front of him. But the precise nature of this wonderful phenomenon is at present beyond man's power to explain.

So much for the largest members of our circle—largest though lightest—members some, visitors others. Now we turn to the smallest.

CHAPTER VIII.

LITTLE SERVANTS.

"All are Thy servants."—PSA. 119:91.

IF you walk out any night after dark, and watch the bright stars shining in a clear sky—shining as they have done for ages past—you will probably see now and then a bright point of light suddenly appear, dart along a little distance, and as suddenly vanish.

That which you have seen was not the beginning of a story, but in ninety-nine cases out of a hundred it was the end of a story. The little shooting-star was in existence long before you saw him, whirling through space with millions of little companions. But he has left them all and dropped to earth. He is a shooting-star no longer.

If such a journey to the moon as the one described two chapters back were indeed possible, the voyage aloft would hardly be so easily and safely performed as is there taken for granted.

Putting aside the thought of other difficulties,

such as lack of conveyance and lack of air, there would be the danger of passing through a very considerable storm of missiles—a kind of "celestial cannonade"—which, to say the least, would prove very far from agreeable.

It will scarcely be believed what numbers of these shooting-stars or meteors—for they are not really stars at all—constantly fall to the earth. As she travels on her orbit, hurrying along at the rate of nineteen miles each second, she meets them by tens of thousands. They too, like the earth, are journeying round the great centre of our family. But they are so tiny, and the earth by comparison is so immense, that her strong attraction overpowers one after another, drags it from its pathway, and draws it to herself.

And then it falls, flashing like a bright star across the sky, and the little meteor has come to his end. His myriads of companions, hastening still along their heavenly track—for the meteors seem to travel commonly in vast flocks or companies—might, had they sense, mourn in vain for the lost members of their family.

Any one taking the trouble to watch carefully some portion of the sky after dark may expect to see each hour about four to eight of these shooting-stars, except in the months of August and November, when the number is much larger. About six in an hour does not sound a great deal. But that merely means that there have been six in one direction and near enough for you to see. Somebody else watching may have seen six in another direction, and somebody else a few miles away may have seen six more. It is calculated that in the course of every twenty-four hours over twenty millions of meteors fall to earth large enough to be visible as shooting-stars. .

This is rather startling. What if you or I should some day be struck by one or other of these little bodies, darting as they do towards earth with speed swifter than that of a cannon-ball? True, they are not really stars, neither are they really planets. But they must be at least often much larger than a cannon-ball; and a cannon-ball can destroy life.

Over twenty millions every twenty-four hours! Does it not seem singular that we do not see them constantly dropping to the ground?

Doubtless we *should* see them, and feel them too, and dire would be the danger to human life, but for a certain protecting something folded round this earth of ours, to ward off the peril.

That "something" is the earth's atmosphere. But for the thick soft air through which they have to pass, they would fall with violence, and perhaps would do much damage.

As it is we are guarded. The "shooting-star," caught by earth's attraction, drops into her atmosphere, darting with extreme speed. The resistance of the air causes it to become intensely heated, till, with a flash of light, it vanishes into vapor. Meteors are believed to appear at a height of somewhere about seventy miles, and to disappear at a height of about fifty miles. So that in one instant's flash the shooting-star has travelled some twenty miles towards us. Then the light goes out. The little meteor is done for. It falls to earth still, but only as fine dust, sinking harmlessly downward.

Such small celestial bodies, thus travelling, do not always vanish so quickly. Now and again one actually reaches the ground; and if a man were struck by such a stone he would undoubtedly be killed. A mason in France once had a narrow escape from such an accident.

Bodies thus falling are called *meteorites* or *aerolites*. Some are found no bigger than a man's fist, while others much exceed this size. There is one, kept carefully in the British Museum, which weighs three tons and a half; and we hear of another, lying in South America, between seven and eight feet in length.

Such a sky-visitant would be very unwelcome in an English or American town.

We must remember that, whatever size an aerolite may be when it reaches the earth, it must have been far bigger when journeying round the sun, since part of it has been vaporized during its rush through earth's atmosphere.

Occasionally meteors are seen of a much larger type, slower in movement and longer visible. These sometimes burst with a loud explosion. They are then often called bolides or fire-balls.

Early in the present century a meteor visited Normandy. It exploded with a noise like the roll of musketry, scattering thousands of hot stones over a distance of several miles. Happily no one seems to have been injured. Other such falls have taken place from time to time.

Sometimes bright slowly-moving meteors have been seen looking as large as the moon.

If we may liken comets to the *fishes* of the Solar

System—and in their number, their speed, their varying sizes, their diverse motions, they may be fairly so likened—we may perhaps speak of meteors as the *animalculæ* of the Solar System.

For in comparison with the planets they are, in the matter of size, as the animalculæ of our ponds in comparison with human beings. In point of numbers they are countless.

Take a single drop of water from some long-stagnant pond and place it under a powerful microscope. You will find it to be full of life, teeming with tiny animals darting briskly to and fro. The drop of water is in itself a world of living creatures, though the naked eye of man could never discover their existence.

So too with the meteors. There is good reason to believe that the Solar System fairly teems with them. We talk of "wide gaps of empty space" between the planets; but how do we know that there is any such thing as empty space to be found throughout all the sun's domain?

Not only are the meteors themselves countless, a matter easily realized, but the families or systems of meteors appear to be countless also. They, like the systems of Jupiter and Saturn, are each a family within a family—a part of the Solar System, and yet a complete system by themselves. Each circles around the sun, and each consists of millions of these tiny bodies. When I say "tiny" I mean it of course only by comparison with other heavenly bodies. Many among them may possibly be hundreds of feet and even more in diameter, but the greater proportion appear to be much smaller. It is not impossible that multitudes beyond imagination exist so small in size that it is impossible we should ever see them, since their dying flash in the upper regions of our atmosphere would be too faint to reach our sight.

The earth, travelling on her narrow orbit round the sun, crosses the track of about one hundred of these systems or rings. Sometimes she merely touches the edge of a ring, and sometimes she goes into the very thick of a dense shower.

Twice every year, for instance, on the 10th of August and the 11th of November, the earth passes through such a ring, and very many falling stars may be seen on those nights. Numbers of little meteors, dragged from their orbits and entangled in the earth's atmosphere, like a fly caught in a spider's web, give their dying flash and vanish.

It used to be supposed that the August and November meteors belonged to one single system; but now they are believed to be two entirely distinct systems.

Once in every thirty-three years we have a grand display of meteors in November, tens of thousands being visible in one single night. The meteors in that ring have their "year" of thirty-three earthly years, and once in the course of that long year our earth's orbit carries her deep into their midst. In this single November ring there are myriads upon myriads of meteors spreading through millions of miles of space.

Yet this system is but one among many. There is no reason whatever to suppose that the streams of meteors cluster more thickly about the orbit of the earth than in other parts of the Solar System. No doubt the rest of the planets come across quite as many.

Indeed the wonderful rings of Saturn are probably formed entirely of meteors—millions upon millions of them whirling round the planet in a regular orbit-belt, lit up by the rays of the sun.

Also it is believed that the meteor families cluster more and more closely in the near neighborhood of the sun, rushing wildly round him and falling by millions into the ocean of fire upon his surface. It has even been guessed that they may serve in part as fuel to keep up his intense heat.

This we do not know with any certainty. But I have called these little bodies "servants," for though we cannot yet say fully what is their precise use and purpose, we know that a use and purpose they must have. "All are God's servants," whether great or little.

There is a curious cone-shaped light seen sometimes in the west after sunset. It is called the "Zodiacal Light," and men have often been much puzzled to account for it. The shining is soft and dim, only to be seen when the sky is clear and only to be seen in the neighborhood of the sun. This too *may* be caused by reflected light from countless myriads of meteors gathering thickly round the sun.

CHAPTER IX.

NEIGHBORING FAMILIES.

" Lift up your eyes on high, and behold who hath created these things, that bringeth out their host by number: He calleth them all by names, by the greatness of His might, for that He is strong in power; not one faileth." Isa. 40 : 26.

WE have now to take flight in thought far, far beyond the outskirts of our little Solar System. Yes, our *great* Solar System, with its giant sun, its planets, its moons, its comets and meteors, its ceaseless motions, its vast distances—even all this sinks to littleness beside the wider reaches of space which now have to be pictured to our minds.

For our sun, in all his greatness, is only a single star—only one star among other stars—and not by any means one of the largest of the stars.

How many stars are there in the sky?

Look overhead some cloudless night and try to count the brilliant points of light. "Millions" you would most likely give as your idea of their number. Yet you would be wrong, for you do not really perceive so many.

The stars visible to man's naked eye have been mapped and numbered. It is found that from two to three thousand are, as a rule, the utmost ever seen at once, even on a favorable night and with particularly good sight.

But what is actually the full number of the stars?

Two or three thousand overhead. Five or six thousand round the whole world. So much visible to man's unaided eyes.

Ah, but take a telescope and see through it the opening fields of stars beyond stars. Take a stronger telescope, and note how, as you pierce deeper into space, fresh stars beyond fresh stars shine faintly in the measureless distances. Take the most powerful telescope ever made, and again it will be the same story.

There has been a chart or map drawn of known stars in the northern hemisphere—including those visible in telescopes down to a certain magnitude—containing over three hundred thousand. But that is only a part of even what man can see. Sir W. Herschel reckoned that the stars within reach of his telescope, round the whole earth, amounted to twenty millions. With the largest modern telescopes the numbers perhaps amount to something like sixty millions, perhaps to very many more.

Sixty millions of suns! For that is what it really means. Sixty millions of radiant shining heavenly bodies, some the same size as our sun, some larger, perhaps very much larger, some smaller, perhaps very much smaller, but all, or most of them, SUNS. And any of these suns may have, like our own, families of planets travelling round them, enjoying their light and their heat.

We talk about stars of the first, second, and other magnitudes. Stars can be seen without a telescope as low down as the sixth magnitude; after that they become invisible to the naked eye.

This word "magnitude" is rather misleading. "Magnitude" means size, and whatever the real size of the stars may be, they have to our sight no seeming size at all.

So when we speak of different *magnitudes* we really mean different *brightnesses*. The brightest stars are those of the first magnitude, the next brightest those of the second magnitude, and so on. No doubt many a star of the third or fourth magnitude is really much larger than many a star of the first or second magnitude, only being farther away it shines more dimly, or the higher-magnitude star may in itself possess greater natural brilliancy.

Of first-magnitude stars there are altogether about twenty; of second-magnitude stars about sixty-five; of third-magnitude stars nearly two hundred; and so the numbers increase, till of the sixth-magnitude stars we find more than three thousand. These are all that can be commonly seen with the naked eye, amounting to five or six thousand. With telescopes the numbers rise rapidly to tens of thousands, hundreds of thousands, even millions and tens of millions.

For a long while it was found quite impossible to measure the distances of the stars. To this day the distances of not over a hundred among all those tens of millions have been discovered. The difficulty of finding out the distance of the sun was as nothing compared with the difficulty of finding out the distances of the stars.

No base-line sufficient for the purpose could for years be obtained. I must explain slightly what is meant by a "base-line." Suppose you were on the brink of a wide river which you had no means of crossing, though you wished to discover its breadth. Suppose there were on the opposite brink a small tree standing alone. As you stood you would see the tree seeming to lie against a certain part of the country beyond. Then if you moved along your bank some fifty paces you would see the tree still, but it would seem to lie against quite a different part of country beyond.

Now if you had a long piece of string to lay down along the fifty paces you walked, and if two more pieces of string were tied, one from each end of the fifty paces, both meeting at the tree, then the three pieces of string would make one large triangle, and the "fifty paces" would be the base of your triangle.

If you could not cross the river you could not of course tie strings to the tree.

But having found your *base-line* and measured its exact length, and having also found the shape of the two angles at its two ends, by noting the seeming change of the tree's position, it would then be quite easy to find out the distance of the tree. The exact manner in which this calculation is made can hardly be understood without some slight knowledge of a science called trigonometry. The tree's distance being found, the breadth of the river would be known.

This mode of measuring distance was found comparatively easy in the case of the moon.

In the case of the sun there was more difficulty, on account of the sun's greater distance. No base-line of ordinary length would make the sun seem to change his position in the sky in the slightest degree. Nor till the very longest base-line on the earth was tried could the difficulty be overcome. That base-line is no less than eight thousand miles long. One man standing in England, looking at the sun, and another man standing in Australia, looking at the sun, have such a base-line lying between them, straight through the centre of the earth.

In the case of the stars this plan was found useless. So closely has the sky been mapped out, and so exactly is the place of each star known, that the tiniest change would have been at once noticed. Not a star showed the smallest movement. The eight thousand miles of the earth's diameter was a mere point with regard to them.

A bright idea came up. Here was our earth travelling round the sun in an orbit so wide that in the middle of summer she is one hundred and eighty-five millions of miles away from where she is in mid-winter. Would not that make a magnificent base-line? Why not observe a star in summer and observe the same star again in winter, and then calculate its distance?

This too was done. For a long while in vain! The stars showed no signs of change beyond those due to causes already known.

Astronomers persevered, however, and with close and earnest care and improved instruments success at last rewarded their efforts. A few— only a few, but still a few—of those distant suns have submitted to the little measuring line of earth, and their distance has been roughly calculated.

Now what is their distance?

Alpha Centauri, the second star which was attempted with success, is the nearest of all whose distance we know.

You have heard how far the sun is from the earth. The distance of Alpha Centauri is *two hundred and twenty-five thousand times as much.*

Can you picture to yourself that vast reach of

space—a line ninety-two millions of miles long, repeated over and over again two hundred and twenty-five thousand times?

But Alpha Centauri is one of the very nearest. 61 Cygni is five hundred thousand times as far as the sun, and Sirius nearly a million times. Others utterly refuse to show the smallest change of position.

It is with them, as has been said, much the same as if a man were to look at a church steeple, twenty miles distant, out of one pane in a window, and then were to look at it out of the next pane. With the utmost attention he would find no change of position in the steeple. And like the base-line of two glass panes to that steeple, so is the base-line formed by our whole yearly journey to thousands of distant stars. We *might* measure how far away they are, only the longest base-line within our reach is too short for our purpose.

The planet Neptune has a wider orbit than ours. But even his orbit, seen from the greater number of the stars, would shrink to a single point.

After all, how useless to talk of two hundred thousand times ninety-two millions of miles!

What does it mean? We cannot grasp the thought.

Let us look at the matter from another view.

Do you know how fast light travels—this bright light shining round us all day long? Light, so far as we know, does not exist everywhere. It travels to and fro, from the sun to his planets, from the planets to one another, from the sun to the moon, from the moon to the earth, and from the earth to the moon again.

Light takes time to travel. This sounds singular, but it is true. Light cannot pass from place to place in no time.

Light, journeying through space, is invisible. Only when it strikes upon something, whether a solid body or water or air, does it become visible to our eyes. The shining all round us in the daytime is caused by the sunlight being reflected, not only from the ground, but from each separate particle of air. If we had no atmosphere we should see still the bright rays falling on the ground, but the sky above would be black. Yet that black sky would be full of millions of light-rays journeying hither and thither from sun and stars, invisible except where they alight upon something.

The speed of light is far beyond that of an express-train, far beyond that of the swiftest planet. Between two ticks of the clock Mercury has rushed onward thirty-five miles. In the same space storm-flames upon the surface of the sun will sweep over two or three hundred miles. But in the same space a ray of light flashes through one hundred and eighty-six thousand three hundred miles.

One hundred and eighty-six thousand miles! That is the same as to say that during one single instant a ray of light can journey a distance equal to about eight times round and round our whole earth at the equator.

By using this wonderful light-speed as a measurement we gain clearer ideas about the distances of the stars.

A ray of light takes more than eight minutes to pass from the sun to the earth. Look at your watch and note the exact time. See the hand moving slowly through the minutes, and imagine one single ray of light, which has left the sun when first you looked, flashing onward and onward through space, one hundred and eighty-six thousand miles each second. Eight minutes and three-quarters end. The ray falls upon your hand.

In those few minutes it has journeyed ninety-two or ninety-three millions of miles.

So much for the sun's distance. How about the stars?

Alpha Centauri, a bright star seen in the southern hemisphere, is one of our nearest neighbors. Yet each light-gleam which reaches the eye of man from that star left Alpha Centauri four years and four months before. During over four years, from the moment when first it quitted the surface of the blazing sun, it has flashed ceaselessly onward one hundred and eighty-six thousand miles each second, dwindling down with its bright companion-rays from a glare of brilliancy to a slender glimmer of light till it reaches the eye of man.

Four years and four months sounds much side by side with the nine minutes' journey from the sun.

But look at 61 Cygni, distant somewhere about five hundred thousand times as far as the sun. The light of 61 Cygni takes more than seven years to reach the earth.

Look at Sirius, that beautiful star so familiar to us all. The light which reaches you to-night from Sirius left his surface nearly nine years ago.

During the greater part of nine years that bright ray has been speeding onward and onward with ceaseless rapidity, till its vast journey is so far accomplished that it has arrived at our earth.

Four—seven—nine years—at the rate of 186,300 miles a second.

These stars are among the few whose distances can be roughly measured. Little can be known with certainty about stars lying yet farther away. The light of a star of the third magnitude may perhaps take, on an average, fifty or sixty years to travel to us. Each lessening magnitude—more truly, each fainter light-gleam—means longer and longer journeys, except in occasional instances, where a larger and brighter sun is actually farther away than another more small and dim. The sixteenth-magnitude stars, just visible in the largest telescopes, may mean a light-journey from each of tens of thousands of years. But here, indeed, we are feeling our way in the dark. Others yet again may lie beyond in countless numbers, at measureless distances, all unknown to us.

Through long centuries these star light-rays have been journeying onward, fulfilling incessantly their Maker's will. On that distant mission each ray started from the far-off sun which gave it birth — when? There is solemnity in the thought. Hundreds of years ago. Thousands of years ago. Some maybe even tens of thousands of years ago! It carries us out of the little present into the unknown ages of a past eternity.

CHAPTER X.

OUR NEIGHBORS' MOVEMENTS.

" Is not God in the height of heaven? and behold the height of the stars, how high they are!" JOB 22 : 12.

How high! how distant! how mighty! How little we know about them, yet how overwhelming the little we know, and how wonderful must be our minds to be able to know it!

We have now to consider the movements of these distant neighbors—first, their seeming movements ; secondly, their real movements.

I have already spoken about the seeming motions of the stars as a whole, once believed to be real and now known to be only caused by the motions of our earth.

For just as the turning of the earth upon her axis makes the sun seem to rise every morning in the east and to set every evening in the west, so that same continued turning makes the stars seem to rise every evening in the east and to set every morning in the west.

When we speak of the stars as rising in the east we do not mean that they all rise at one point in the east, but that all rise, more or less, in an easterly direction—northeast, east, and southeast. So also with respect to the west. It is to the east and west of the earth as a whole that they rise and set—not merely to the east and west of that particular spot on earth where one man may be standing.

All night long fresh stars are rising and others are setting; and if it were not for the veil of light made by the sunshine in our atmosphere, we should see the same going on all day long as well.

There are some constellations, or groups of stars, always visible at night in our northern hemisphere; and there are some constellations never visible to us, but only seen by people living in the southern hemisphere—in Australia for instance.

There are other constellations which appear in summer and disappear in winter or which appear in winter and disappear in summer. This change is caused by our earth's journey round the sun, or, as in common language we express it, by the sun's journey round the earth. It is not that the constellations have altered their place in the heavens with respect to the other constellations: it is merely that the sun has so altered his position in the heavens that the group of stars which a short time ago were above the horizon with him by day are now above the horizon without him by night.

Mention has been a good many times made of the axis of the earth ending in the north and south poles.

If this axis were carried straight onward through space, a long slender pole passing upwards into the sky without any bend, from the North Pole in one direction and from the South Pole in the other, this would be the Pole of the heavens. The places of the stars in the sky are counted as "so many degrees" from the North and South Celestial Poles—just as the places of towns on earth are counted as "so many degrees" from the North and South Poles of earth. There are atlases of the sky made as well as atlases of the earth.

The constellation of the Great Bear is known to all who have ever used their eyes at all to watch the heavens. Almost equally well known are the two bright stars in this constellation named the Pointers, because, taken together, they point in nearly a straight line to a certain important star

Our Neighbors' Movements. 121

in the end of the Little Bear's tail, not very distant.

This star, important less from its brightness than from its position, lies close to that very spot in the heavens where the celestial North Pole passes. It is called the Pole-star. Night after night through the year it there remains, all but motionless, never going below the horizon for us in the northern hemisphere or northern half of the earth, never rising above the horizon for those in the southern hemisphere. It shines ever softly and steadily in its fixed position. If you travel further south, the Pole-star sinks downward towards the horizon. If you travel further north, the Pole-star rises higher above the horizon. If you were at the North Pole, you would see the Pole-star exactly overhead.

Very near the Pole-star is the constellation of the Great Bear, with Cassiopeia nearly opposite on the other side of the Little Bear, and other groups between the two completing the circle. These constellations do not, to us who live in England, rise or set, for they simply move in a circle round and round the Pole-star, never going below the horizon. All day and all night long this circling movement continues, though only visible at night. It is caused entirely by the earth's own motion on her axis.

Lower down, or rather farther off from the Pole-star, comes another ring of constellations. These in just the same manner appear to travel round and round the Pole-star. But being farther away, each dips in turn below the horizon—or, as we call it, each sets and rises again. And by the time we come to yet another circle of leading constellations we reach those which are so far affected by the earth's yearly journey as to be only visible through certain months and to be hidden during other months.

If we could stand exactly at the North Pole during part of its six months' night, we should see the Pole-star just overhead and all the constellations circling round it once in every twenty-four hours. Those nearest would move slowly, in a small ring. Those farthest, and lowest down, would in the same length of time sweep round the whole horizon. But the stars would not there seem to rise or set.

If we were standing at the South Pole we should see exactly the same kind of seeming movement, only with altogether a different set of stars.

If we were standing on the Equator at night we should see the rising and setting very plainly. The whole mass of stars would appear to rise regularly and evenly in an easterly direction, to pass steadily across the sky, each taking its own straightforward path, and to set in a westerly direction.

We in England, placed between the Pole and the Equator, see a mixture of these two motions. Some stars seem to circle round and round, as all would do if we stood at the North Pole. Some stars seem to rise and set, as all would do if we stood at the Equator.

So much for the seeming movements of the stars.

But now about their real movements. Are the stars fixed or are they not?

These seeming daily and yearly motions do not affect the question, being merely caused by our own motions. Trees and hedges may appear to move as we rush past them in a train, yet they are really fixed.

During a long while, after it was found out that the quick daily movements of all the stars in company were merely apparent, men believed that they really had no "proper motions," that is, no movements of their own.

For century after century the constellations remain the same. Hundreds of years ago the seven chief stars of the Great Bear shone in company as they shine now. Who could suppose that each one of those seven stars is hurrying on its path through space with a speed exceeding far that of the swiftest express-train? Yet so it is. Hundreds of years ago the grand group of Orion, with belt and sword, gleamed brilliantly night by night as it gleams in these days, and Cassiopeia had her W-form, and Hercules and Draco and Andromeda were shaped as they are shaped still. Who would imagine that through those hundreds of years each star of those different constellations was hastening with more or less of speed along its heavenly road? Yet so it is.

But if the stars are thus rapidly moving in all directions, how is it that we do not *see* them move? How is it that night after night, year after year, century after century, even thousand years after thousand years, the shapes of the constellations remain unaltered?

Suppose you and I were standing on the seashore together, watching the movements of scores of seacraft, little boats and large boats, steamers, yachts, and ships. Suppose we stood through a full quarter of an hour looking on. Some might move, it is true, very slowly, yet their movements in every case would plainly be seen. There could be no possibility of mistaking the fact or of supposing them to be "fixed."

Just so we see the nearer planets move. Little danger of our supposing them to be "fixed stars."

In the matter of the stars themselves we must carry our illustration farther.

Come with me up to the top of that lofty hill on the border of the sea and let us look from the cliff. We see still the movements among boats and smacks, yachts and steamers, only the increased distance makes the movements seem slower.

But our view is widened. Look on the far horizon and see three distant dots, which we know to be ships — one and two close together and a third a little way off, making a small constellation of vessels. Watch them steadily for a quarter of an hour. You will detect no movement, no increased distance or nearness between any two of the three. The group remains unchanged.

Are they really moving? Of course they are, more or less rapidly, probably with differing speed and in different directions. But at so great a distance one quarter of an hour is not long enough for their motions to become visible to the naked eye. If we could watch longer, say for two or three hours, ah, that would make all the difference! If only we could watch longer! But the hundreds and even thousands of years during which men have watched the stars, sink, at our vast distance, into no more than one quarter of an hour spent in watching the far-off ships from the hill-top. The motions cannot be detected. In ten thousand years you might see something. In fifty thousand years you might see much. But four or five thousand years are not sufficient.

One other mode there is by means of which the movements of the ships on the horizon might be made plain. Suppose you had no more than the quarter of an hour to spare, but suppose you had at your command a powerful telescope. Then you may practically bring the ships nearer, and by magnifying the small slow distant motions you may make them, as it were, larger, quicker, more easy to see.

Telescopes will do this for us, likewise, in the

matter of the stars. By means of telescopes, with the assistance of careful watching and of close calculation, it has been found that the stars are really moving quickly, each one in his own pathway. The very speed of some of them has been measured.

One fair star, Capella, speeds away from us at about the rate of one thousand miles each minute, or some sixteen or seventeen miles each second; and the pale Pole-star, friend of mariners, is on the other hand advancing towards us no less rapidly. The brilliant Sirius recedes at a pace of over twenty-six miles per second. There is good reason to believe that every star seen in the heavens, every star visible in most powerful telescopes, to the tune of millions, is incessantly hastening onward.

Hastening whither?

God knows! We do not. Each radiant sun has its goal, but where that goal may be we cannot tell. Each doubtless has its own pathway around some distant centre—the centre of that Sun-cluster or Steller System or Starry Universe to which it belongs. Each follows its own pathway, controlled and regulated by the united attractions of all other suns belonging to the same Star-System: even as the planets of the Solar System are controlled by the central sun and influenced by each brother-planet.

Yet it may be that not all the stars in our Stellar System are so controlled. Some few we know travelling at so vast a rate of speed as to have earned the title—whether rightly or wrongly—of "Runaway Stars." Sixteen miles, twenty miles, twenty-six miles per second sounds rapid enough; but these rates are eclipsed. A certain dim double star, 61 Cygni, journeys at least twice as fast as Capella, or some thirty-six miles a second. This may still be a controlled speed, and 61 Cygni can still be counted a permanent member of our System, subject to the restraint of brother-stars.

But when we reach a certain degree of speed, then, it is conjectured, the power of this control *may* fail. One faint star, known as *Groombridge* 1830, flies at the rate of over two hundred miles each second; and Arcturus, a brilliant star of the first magnitude, is believed to whirl along at the pace of three hundred and seventy-five miles each second. A star pursuing such a headlong career as either of the two last named, may—not improbably—defy the combined control of all the other stars. Such stars are, perhaps, not members of our Stellar System, but passing visitors, entering from one direction, passing away in another direction.

The thought of such bright suns, travelling through long ages from far far distant Star-Systems, entering our own, and during ages more journeying among *our* innumerable stars, only, in ages yet future, to pass away into depths of space beyond, perhaps to join some other starry universe—could aught be more full of suggestiveness as to the meaning of Infinity? The life of mankind sinks to a point by comparison, the life of a man to the flutter of a May-fly. And whether or no the story of Runaway Stars be as thus pictured, such distances *do* exist!

If all the stars are moving, what of our sun? Our sun is a star!

And our sun also is moving. He is pressing onward in a wide sweep through space, bearing along with him his whole family—planets, satellites, comets, meteors—round or towards some far-off centre, at a probable rate of sixteen or eighteen miles each second.

And where are we going?

This has been in part discovered.

If you and I were driving through a forest of trees, we should see trees on each side of us seeming to move backward, while behind they would close together and in front they would open out.

Astronomers—and first among them William Herschel—reasoned that if our Solar System were really in motion we ought to be able to see these changes among the stars. And some such changes have become visible through careful watching— not so much those ahead and behind as those at the sides.

It is not actually so simple a matter as looking at the trees in a forest, because the trees would be at rest, whereas each star has his own particular real motion as well as his seeming change of place caused by our sun's motion. It is more like moving in a small steamer at sea among hundreds of other craft, each of which is going on its own way, at the same time that all on either side seem to move backward because we are moving forward.

So each movement had to be noted, and the real motions had to be separated from the seeming backward drift of stars to the right and left of the sun's pathway. The result of all this is that the sun, with his planets, is found to be hastening towards a certain far-off constellation named Hercules.

Where the sun and his planets will journey in future ages no living man can say. Indeed, though it is a question which does not lack interest to a thoughtful mind, yet there are numberless other questions about centuries near at hand which concern man far more nearly. The history of the universe and the history of this earth of ours, in its close and wonderful connection with God and with Christ our Redeemer, must have advanced many broad stages before our sun and his attendant planets can have travelled so far that any change will be apparent in the shape of the star-constellations which spangle our sky.

END OF PART I.

PART II.

Fig. 1.—Total Eclipse of the Sun.

The eclipse is total upon that portion of the Earth upon which the cone of dark shadow falls; and partial in the region of the lighter shade, or penumbra.

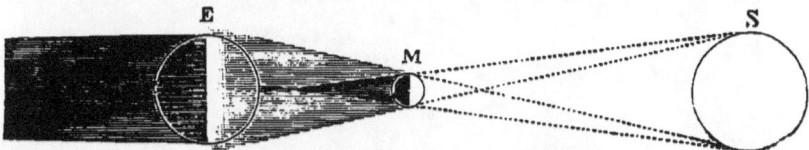

Fig. 2.—Annular Eclipse of the Sun.

The eclipse appears annular upon that portion of the Earth upon which the *reversed* dark cone of shadow falls.

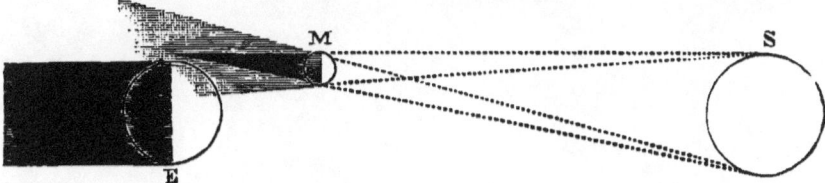

Fig. 3.—Partial Eclipse of the Sun.

Fig. 4.—Total Eclipse of the Moon.

Fig. 5.—Partial Eclipse of the Moon.

CHAPTER I.

MORE ABOUT THE SOLAR SYSTEM.

"By the word of the Lord were the heavens made, and all the host of them by the breath of His mouth. . . . For He spake, and it was done; He commanded, and it stood fast."—PSA. 33:6, 9.

WE have now reached a point where it ought not to be difficult for us to picture to ourselves with something of vividness the general outlines of the Solar System.

A while ago this Solar System was a very simple matter in the eyes of astronomers. There was the great sun fixed in the centre, with seven planets circling round him—seven of course, it was said, since seven was the perfect number—and a few moons keeping pace with some of the planets, and an occasional comet, and a vast amount of black empty space.

But astronomers now begin to understand better the wonderful richness of the System as a whole, the immense variety of the bodies contained in it, the perpetual rush and stir and whirl of life in every part. Certainly there is no such thing as dull stagnation throughout the family.

First we have the great blazing central sun; not a sun at rest as regards the stars, but practically at rest as regards his own system, of which he is always head and centre. Then come the four smaller planets, rapidly whirling round him, all journeying in the same direction and all having their oval pathways lying on nearly the same flat plane in space. Then the broad belt of busy little planetoids. Then the four giant planets, Jupiter nearly five times as far as our earth from the sun, Saturn nearly twice as far as Jupiter, Uranus nearly twice as far as Saturn, Neptune as far from Uranus as Uranus from Saturn, all keeping on very nearly the same level as the four inner planets.

And between and about these principal members of the system, with their accompanying moons, we have thousands of comets flashing hither and thither with long radiant trains; and myriads of meteorites, gathered often into dense vast herds or families, but also scattered thickly throughout every part of the system, each tiny ball reflecting the sun's rays with its little glimmer of light.

Broad reaches of black and empty space! Where are they? Perhaps nowhere. We are very apt, in our ignorance, to imagine that where we see nothing there must of necessity be nothing.

But for aught we know the whole Solar System, not to speak of sky-depths lying beyond, may gleam with reflecting-bodies great and small, from the great Jupiter down to the fine diamond-dust of countless meteorites. In this earth of ours we find no emptiness. Closer and closer examination with the microscope only shows tinier and yet tinier wonders of form and life, each perfect in finish.

Not of form only, but of *life*. How about that matter as regards the Solar System? Is our little world the one only spot in God's great universe which teems with life? Are all other worlds mere barren empty wastes?

Surely not all! So much at least we may safely surmise. Life of one kind or another probably either has been, or is, or will be at some future period, upon at all events *some* of the countless worlds existent in the universe.

On the other hand we must remember that surmise is not certainty: and that we know, and can know, little or nothing about the matter. The many worlds of the Solar System may have been created for some widely different purpose, beyond our present power to fathom or even to imagine. It is not necessary that every world should be formed for the one object of supporting life. It may even be, as has been suggested by one of the greatest astronomers of our day, that the other planets are but as "chips struck from the block" and flung aside, in the making of our world — a world by no means prominent in size or in weight or in radiance, and yet perhaps the greatest in the universe, because of the marvellous development of mind, and because of the extraordinary history of mankind, connected with it.

The same reasoning may be used for the stars— those millions of suns lying beyond reach of man's unassisted eyes. Are they formed in vain? Do their beams pour uselessly into space, carrying light and warmth and life-giving power to nothing? But here again how little we know! We stand on the shore of a boundless ocean, able to see but a little way, able to understand not a hundredth part of what we see! That neither worlds nor suns are objectless or useless may be safely asserted, within certain limits; but what the especial object and use of each bright orb may be we may not venture to assert—we can hardly venture even to conjecture.

I have spoken of the probable fulness of the Solar System as a whole. We are so apt to think of things merely as we see them with our short sight that it is well sometimes to try to realize them as they actually are.

Picture to yourself the great central sun pouring out in every direction his burning rays of light. A goodly abundance of them fall on our earth, yet the whole amount of light and heat received over the whole surface of this world is only the two-thousand-millionth part of the enormous amount which he lavishly pours abroad into space.

How much of that whole is wasted? None; though, like other divine gifts in nature, light is given with a kingly profusion which knows no bounds. Millions of rays are needed for the lighting and nourishing and warming of our companion-planets, while others are caught up by passing comets and myriads flash upon swift small meteors. Of the rays so used many pass onward into the vast depths beyond our system, and dwindle down into dim star-like shining till they reach the far-off brother-stars of our sun.

Have they aught to do there? We cannot tell. We do not know how far the sun's influence reaches. As head and centre he reigns only in his own system. As a star among stars, a peer among his equals, he may, for aught we can tell, have other work to do.

In an early chapter mention was made of the earth's three motions, two only being explained.

First, she spins ceaselessly upon her axis. So does the sun and so do the planets.

Secondly, she travels ceaselessly round and round the sun in her fixed orbit. So does each one of the planets.

Thirdly, she journeys ceaselessly onward through space with the sun. So also do the rest of the planets.

These last two movements, thought of together, make the earth's pathway rather perplexing at first sight. We talk of her orbit being an ellipse or oval; but how can it be an ellipse if she is always advancing in one direction?

The truth is the earth's orbit is and is not an ellipse. As regards her yearly journey around the sun, roughly speaking we may call it an ellipse; as regards her movement in space, it certainly is not an ellipse.

Think of the Solar System with the orbits of all

More About the Solar System. 141

the planets as lying *nearly flat*—in the manner that hoops might be laid upon a table, one within another. The asteroids, comets, and meteorites do not keep to the same level, but their light weight makes the matter of small importance.

Having imagined the sun thus in the centre of a large table—a small ball, with several tiny balls travelling round him on the table at different distances — suppose the sun to rise slowly upwards, not directly up, but in a sharp slant, the whole body of planets continuing to travel round and at the same time rising steadily with him.

By carefully considering this double movement you will see that the real motion of the earth—as also of each of the planets—is not a going round on a flat surface to the same point from which she started, but is a corkscrew-like winding round and round upwards through space. Yet as regards the central sun the shape of the orbit comes very near being an ellipse, if calculated simply by the earth's distance from him at each point in turn of her pathway through the year.

An illustration may help to explain this. On

the deck of a moving vessel you see a little boy walking steadily round and round the mast. Now is that child moving in a circle or is he not?

Yes, he is. No, he is not. He walks in a circle as regards the position of the mast, which remains always the centre of his pathway. But his movement *in space* is never a circle, since he constantly advances and does not once return to his starting-point. You see how the two facts are possible side by side. Being carried forward by the ship, with no effort of his own, the forward motion does not interfere with the circling motion. Each is performed independently of the other.

It is the same with the earth and the planets. The sun, by force of his mighty attraction, bears them along wherever he goes, no exertion on their part, so to speak, being needed. That motion does not in the least interfere with their steady circling round the sun.

Just as—to use another illustration—the earth, turning on her axis, bears through space a man standing on the equator at the rate of one thousand miles an hour. But this uniform movement, unfelt by himself, does not prevent his walking

SUN FLAMES.
Seen during a Total Eclipse—Telescopic View.

backwards or forwards or in circles as much as he will.

So also a bird in the air is unconsciously borne along with the atmosphere, yet his freedom to wheel in circles for any length of time is untouched.

A few words about the orbits of the planets.

I have more than once remarked that these pathways are in shape not circles, but ellipses. A circle is a line drawn in the shape of a ring, every part of which is at exactly the same distance from the centre-point or focus. But an ellipse, instead of being like a circle perfectly round, is oval in shape; and instead of having only one focus, it has two foci, neither being exactly in the centre. Foci is the plural word for focus. If an ellipse is only slightly oval—or slightly *elliptical*—the two foci are near together. The more oval or *eccentric* the ellipse, the farther apart are the two foci.

You may draw a circle in this manner. Lay a sheet of white paper on a board and fix a pin through the paper into the board. Then pass a loop of thread—say an inch or an inch and a half in length—round the pin, and also round a pencil, which you hold. Trace a line with the pencil, keeping the loop tight, so that the distance of your line from the pin will be always equal, and when it joins you have a circle. The pin in the centre is the focus of the circle.

To draw an ellipse you must fix *two* pins. Let them be about half an inch apart; pass a loop over both of them, and again placing a pencil-point within the loop, again trace a line carefully all round, keeping the thread drawn tight. This time an oval instead of a circle will appear. By putting the pins nearer together or farther apart you may vary as you will the shape of the ellipse.

In the orbits of the earth and the planets, all of which are ellipses in shape, the sun is not placed in the exact centre, but in one of the two foci, the second being empty. So at one time of the year the planet is nearer to the sun than at another time. Our earth is no less than three millions of miles nearer in winter than she is in summer—speaking of the winter and summer of the northern hemisphere. Three millions of miles is so tiny a piece out of ninety-two millions of miles that it makes little or no difference in our feelings of heat or cold.

More About the Solar System. 145

The orbits of some comets are ellipses also, but ellipses often so enormously lengthened out that the two foci are almost—if one may so speak—at the two *ends* of the oval. To draw a good comet orbit you must fix the two pins on your paper some five or six inches apart, with a loop of thread just large enough to slip over them both and to allow the pencil to pass round them. When your ellipse is drawn you must picture the sun in the place of one of the two pins, and you will see how in their pathways the comets at one time pass very near the sun and at another time travel very far away from him.

It is generally found in families not only that the parent or head of the family has great influence over all the members, but that each member has influence over each other member. Brother influences brother, and sister influences sister.

This, too, we find in the Solar System.

Not only does the sun by his powerful attraction bind the whole family together, but each member of the family attracts each other member.

True, the force of the sun's attraction is overpowering in amount compared with others. The sun attracts the planets and the planets attract the sun; but their feeble pulling is quite lost in the display of his tremendous strength.

Among themselves we see the power more plainly. The earth attracts the moon, keeping her in constant close attendance; and the moon attracts the earth, causing a slight movement on her part, and also causing the tides of the sea. Each planet has more or less power to hinder or help forward his nearest brother-planet. For instance, when Jupiter on his orbit draws near the slower Saturn on his orbit, Saturn's attraction pulls him on and makes him move faster than usual; but as soon as he gets ahead of Saturn, then the same attraction pulls him back and makes him go more slowly than usual. Jupiter has the same influence over Saturn; and so also have Saturn and Uranus over one another, or Uranus and Neptune.

In early days astronomers were often greatly puzzled by these quickened and slackened movements, which could not be explained. Now the "perturbations" of the planets, as they are called, are understood and allowed for in all calculations.

Indeed it is by means of this very attraction

that the planets have actually been weighed, somewhat in the same manner—to illustrate it roughly—as one might test the weight of two strong children, by setting each to pull against the other and seeing which could pull the hardest.

What a wonderful difference we find in this picture of the Solar System as we now know it to be from the old-world notion of our earth as the centre of the universe!

When we think of all the planets, and of the magnificent sun, when we pass onward in imagination through space, and find our sun himself merely one twinkling star amid the myriads of twinkling stars scattered broadcast through the heavens, while planets and comets have sunk to nothing in the far distance, then indeed we begin to realize the unutterable might of God's power! Why, our earth and all that it contains may be regarded as but one grain of dust in the wide universe.

And yet—

Is there "great" and is there "little" intrinsically? We do not know. Probably not, as we see either. The might of God is no less shown in the exquisite finish of a flower or an insect than in the fierce radiance of the stars. And, little though our earth may be, it was not too small or too unimportant for the kingly Creator himself to dwell here for over thirty years and here to redeem with his life the life of man. For not even in one corner of his boundless universe might evil be left to reign unconquered.

THE MOON.

CHAPTER II.

MORE ABOUT THE SUN.

"He is wise in heart and mighty in strength; ... which doeth great things past finding out; yea, and wonders without number."—JOB 9:4, 10.

NOT among the least of the wondrous things of creation are the tremendous disturbances taking place upon the surface of the sun — that raging roaring sea of flame.

A good many explanations have been from time to time offered as to the dark spots seen to move across the face of the sun. Some one or more of these explanations may be true, but a great deal of uncertainty still exists.

A sun-spot does not commonly consist of merely one black patch. There is the dark centre called the *umbra* — plural, *umbræ*. There is the grayish part surrounding the umbra, called the *penumbra*. Also, in the centre of the umbra, there is sometimes observable an intensely black spot called the *nucleus*.

Sometimes a spot is made up of nucleus and

umbra alone, without any penumbra. Sometimes it is made of penumbra alone, without any umbra. Sometimes in one spot there are several umbræ, with the gray penumbra round the whole and gray bridges dividing the umbræ.

The enormous size of these spots has been already described in an earlier chapter. Fifty thousand to one hundred thousand miles across is nothing unusual. In the year 1873 a spot was seen five times as large as the whole land and water surface of the earth.

One explanation proposed was that the sun might be a cool body covered over with different envelopes or dense layers of cloudy form, one above another. The inside envelope, or as some say the inside atmosphere, would then be thick and dull-colored, protecting the solid globe within, and reflecting light, but having none of its own. The next envelope would be one mass of raging burning gases—the *photosphere*, in fact. The outer envelope would be a transparent surrounding atmosphere lighted up by the sea of fire within. A sun-spot would then consist of the tearing open of one or more of these envelopes so as to give glimpses of the gray inner atmosphere, or even of the dark cool globe at the centre.

There can be little truth in this explanation, since the notion of a cool and dark body within is now pretty well given up. The apparent blackness of a spot-nucleus does not prove actual blackness or absence of heat. A piece of white-hot iron held up against the sun looks black; and it may be merely the contrast of the glowing photosphere which makes the nucleus seem so dark. It is even believed that the blackest parts may be the most intensely hot of all.

Another proposed explanation was of dark clouds floating in the sun's atmosphere. But since the whole photosphere is now believed to consist of *bright* clouds floating in the solar atmosphere, this theory too has gone to the wall.

Much doubt is felt whether the spots really are depressions at all, although they certainly have often the appearance of a hollow and cave-like form.

Beside apparent changes of shape, caused by changes of position as the spots travel across the sun's face, there are also real changes constantly taking place. Although the spots often keep their general outlines long enough to be watched across the face of the sun, and even to be known again after spending nearly a fortnight hidden on the other side, still they are far from being fixed in form.

The alterations are at times not only very great but very rapid. Sometimes in a single hour of watching an astronomer can see marked movement going on—as you or I might in an hour observe movements slowly taking place in a high layer of clouds. For movement to show at all in one hour, at so immense a distance, proves that the actual rate of motion must be very great.

Whatever else we know or do not know about the spots, it seems probable that they are caused by some description of cyclones on the sun's surface. A cyclone is a fierce hurricane of wind blowing round and round in a circle—a whirlwind on a large scale.

These sun-cyclones must indeed be of terrific force and extent, viewed side by side with anything that we see on our earth. The speed of movement perceived in some spots has been calculated to amount to no less than one and two hundred miles or more each second. The most violent hurricanes of earth sink by comparison into nothing.

Sometimes the storms or outbursts come in

the shape of a bright spot instead of a dark one.

Two astronomers were one day watching the sun from two different observatories when they saw such an event take place. An intense and dazzling spot of light burst out upon the surface of the sun—so intense, so dazzling, as to stand quite apart from the radiant photosphere. To one astronomer it looked like a single spot, while the other saw two spots close together. In about a minute the light grew more dim, and in five minutes all was over. But in those five minutes the spot or spots had travelled a distance of thirty-five thousand miles.

It was a notable matter that the *magnets* on earth—those delicate little needles which point so steadily and perseveringly towards the North Pole—seemed to be strongly agitated by the distant solar outburst.

This brings us to another interesting fact.

The spots on the sun are not always the same in number. Sometimes they are many, sometimes they are few. Long and close watching has made it clear that they pass through a regular *order* of changes, some years of many spots being followed by other years of less and less spots; then some years of very few spots being followed by other years of more and more spots, decrease and increase being seemingly regular and alternate.

This turn or *cycle* of changes, from more to less and then from less to more again, is found to run its course about once in every eleven years, with some variations.

Now it has long been known that the magnetic needle goes through curious variations. Though we speak of it as pointing always north, yet it does not always so point exactly.

Every day the needle is found to make certain tiny delicate motions, as if faintly struggling to follow the daily movements of the sun—just a little towards the east or just a little towards the west. These tiny motions, having been long watched and measured, were found to go through a regular course of changes—some years more and some years less, waxing and waning by turns. It was discovered that the course of changes from more to less, and from less to more again, took place in about eleven years.

These two things, you see, were quite independent of one another. Those who watched the sun-spots were not thinking of the magnets, and those who watched the magnets were not thinking of the sun-spots. But somebody did at last happen to think of both together. He was laughed at, yet he took the trouble carefully to compare the two.

And, very remarkably, he found that these two periods in the main agreed—the eleven years of alternate changes in the number of sun-spots and the eleven years of alternate changes in the movements of the magnetic needle. When the spots are most the needle moves most. When the spots are least the needle moves least.

So much we know. But to explain the why and the wherefore is beyond our power.

There is a very singular appearance seen upon the sun which must not be passed over without mention.

Some astronomers speak of the whole surface as being *mottled* all over with a curious rough look when examined through a powerful telescope.

This "mottling" is described by various observers in various ways. One speaks of "luminous spots shaped like rice-grains;" another of "luminous spots resembling strokes made with a camel's hair pencil;" another of "luminous objects or granules;" others of "multitudes of leaves," "nodules," "crystalline shapes," "leaves or scales crossing one another in all directions, like what are called spills in the game of spillikins." They have also been pictured as "certain luminous objects of an exceedingly definite shape and general uniformity of size, whose form is that of the oblong leaves of a willow-tree. These cover the whole disc of the sun, excepting the space occupied by the spots, in countless millions, and lie crossing each other in every imaginable direction."

In size they are said to be about one thousand miles long by two or three hundred broad, but they vary a good deal. Where there is a spot the willow-leaves at its edge are said to point pretty regularly towards the centre.

"It would appear," writes astronomer-royal Ball, alluding to these objects, "as if the luminous surface of the sun was composed of *intensely bright clouds* suspended in a darker atmosphere. Some observers have thought that these floating objects are, occasionally at all events, of a characteristic size and shape, variously known as 'willow leaves' or 'rice granules!'"

We have next to think a little more about the edge or limb of the sun and the stormy flames and outbursts there seen.

Until of late years the only time for observing such appearances was during a total eclipse of the sun. Now, however, by means of the "spectroscope" it has been found possible to take observations when no eclipse is going on.

A few words of explanation as to eclipses of the sun seem needful before going farther.

An eclipse of the sun is caused simply by the round body of the moon passing exactly between the sun and the earth so as to hide the sun from us.

Let there be a candle on the table, while you stand near. The rays of light from the candle fall upon your face. Now move slowly, to and fro, a round ball between you and the candle. So long as it is not precisely in the line between—so long as it is a little higher or a little lower or a little to one side—then you can see the flame. Once let the ball come just between the light and your eyes, and you see it no more. In other words the candle-flame is eclipsed by the ball.

It may seem curious at first sight that the moon, which is so very small compared with the sun, should have power to cover the sun. But remember the difference of the distances. The sun is very far and the moon is very near. Any small object very near will easily hide from your sight a large object at a considerable distance. You may hold up a shilling-piece at arm's length and make it cover from sight a man, or even a house, if the latter be far enough away.

The sun at a distance of ninety-two millions of miles, and the moon at a distance of two hundred and forty thousand miles, have to our vision the same seeming size. So, when the moon glides between, her round face just about covers the sun's round face.

If the moon were travelling exactly in the same plane as the plane of the earth's orbit, an eclipse would be a very common affair indeed. But the plane of the moon's orbit being not quite the same as the plane of the earth's orbit, she passes sometimes a little above and sometimes a little below the exact spot where she would hide the sun's rays from us. Now and then, at certain intervals, she goes just between. And so well is the moon's path in the heavens understood that astronomers can tell us long years beforehand in what day and at what hour an eclipse is certain to take place.

An eclipse of the sun is sometimes partial,

sometimes total, sometimes annular. In a partial eclipse the moon does indeed pass between, but only so as to hide from us *part* of the sun. She is a little too low or a little too high to cover his face. In a total eclipse the moon covers the sun completely, so that for a few minutes the bright photosphere seems blotted out from the heavens, a black round body surrounded by light taking its place. In an annular eclipse the moon in like manner crosses the sun, but does not succeed in covering him entirely, a rim of bright photosphere showing round the black moon. For in an annular eclipse the moon, being a little farther away from the earth than at the time of a total eclipse, has too small a disc quite to hide the sun's disc.

The blackness of the moon during an eclipse is caused by the fact that her bright side is turned towards the sun and her dark side towards us. An eclipse of the sun can only take place at new moon,* never at full moon. At her full the moon

---

* A peculiar interest belongs to this fact. For while our Lord hung upon the cross there came a strange and mysterious darkness over the whole land for three hours. Now it might be supposed that this darkness was caused by an eclipse of the sun—*but* no such eclipse can ever take place except at new moon,

is outside the earth's orbit, away from the sun, and cannot by any possibility pass between.

Eclipses of the sun in olden days were a source of wide-spread awe and terror. Many wild tales and wilder explanations were invented to account for them. Now that we better understand their meaning, fear need no longer be felt, though the strange darkness and the singular coloring of sky and air must always be exceedingly impressive.

The following description of the total eclipse of 1860 will be found interesting. It was observed by several astronomers who stationed themselves at various places in the north of Spain—Mr. Airy, the Astronomer-Royal, at Pobes; M. Bruhns of Leipzic, near Tarragona; Mr. Lowe, near Santander. Mr. Lowe writes:

"Before totality commenced the colors in the sky and on the hills were magnificent beyond all description. The clear sky in the north assumed a deep indigo color, while in the west the horizon was first black like night. In the east the clear sky was very pale blue, with orange and red like sunrise. On the shadow sweeping across, the deep blue in the north changed like magic to

nor for more than six minutes; and the Feast of the Passover, when He was crucified, always took place at the full moon.

pale sunrise tints of orange and red, while the sunrise appearance in the east had changed to indigo. The darkness was great; the countenances of men were of a livid pink. The Spaniards lay down, and their children screamed with fear; fowls hastened to roost, ducks clustered together, pigeons dashed against the houses, flowers closed, many butterflies flew as if drunk and at last disappeared. The air became very humid, so much so that the grass felt to one of the observers as if recently rained upon."*

* From "Descriptive Astronomy," by G. F. Chambers.

CHAPTER III.

YET MORE ABOUT THE SUN.

"O Lord, how great are thy works!" Psa. 92 : 5.

Having seen something of storms taking place on the sun's photosphere, we must next give our attention to storms taking place at his edge. But it should be remembered that the said edge, far from being a mere rim to a flat surface, is a kind of horizon-line—is in fact just that part of the photosphere which is passing out of or coming under our sight. The surface there is in kind the same as the surface of the broad disc facing us. In watching outbursts at the edge of the sun we have a side-view instead of a bird's-eye view.

In the year 1871 an American astronomer, Prof. Young, was looking at a large hydrogen cloud on the edge of the sun. When I speak of a "cloud" it must not be supposed that anything like a damp foggy earthly cloud is meant. This solar cloud was a huge mass of glowing gas about one hundred thousand miles long, rising to a height of fifty thousand miles from the sun's surface and appearing to rest on pillars of fire.

The professor, while watching, was called away for half an hour. He came back expecting to find things much as he had left them. Instead of this a startling change had taken place. The whole mass of crimson fire seemed to have been actually "blown to shreds" by some tremendous outburst from below. In place of the motionless cloud were masses of scattered fire, each from about four thousand to fourteen thousand miles long and a thousand miles wide.

As the professor gazed, these "bits" of broken cloud rose rapidly upwards, away from the surface of the sun. When I say "rapidly," I mean that the real movement, which the professor could calculate, was rapid. The seeming movements were of course slow and over a small space. The actual motions were not tardy, for in ten minutes these huge fiery cloud-pieces rushed upwards to a height of two hundred thousand miles from the edge of the sun, moving at a rate of at least one hundred and sixty-seven miles each second. Gradually they faded away.

But what caused this sudden change?

Just before the professor was interrupted he had noticed a curious little brilliant lump—a sort of suspicious thunder-cloud appearance—below the quiet bright cloud. And after this tremendous shattering the little bright lump rose upwards into a huge mass of rolling flame, reaching like a pyramid to a height of fifty thousand miles. In the course of a few minutes these enormous flames could be seen to move and bend and to curl over their gigantic tips. But they did not last long. At half-past twelve the professor had been called away; by half-past two the rolling flames completely vanished.

Now, whatever may be the full explanation of this sight, there is no doubt that on that day was observed from earth a gigantic outburst, compared with which our greatest volcanoes are like the sputtering of a farthing dip beside a roaring furnace. The awful force and extent of such a solar eruption are more than we can possibly picture to ourselves. At our distance we may catch a faint glimpse of what is going on and calculate speed of movement. But vividly to realize the actual terrific grandeur of what took place is past our power.

Possibly this was much the same kind of outburst as that seen by the two English astronomers; only theirs was a bird's-eye view, as it were, looking down on the top of the sight, while the profes-

sor had a side-view, certainly much the best for observation.

It does not follow from what he saw that the eruption must have taken place exactly at the "edge" of the sun. Probably it happened near the edge. All he could say was that the flames rose fifty thousand miles, and the pieces of cloud were carried two hundred thousand miles, away from the edge. The eruption may have begun on the other side of the sun, at any distance from the horizon-edge where it first became visible to earthly eyes.

Also, while the professor found that the shattered cloudlets moved at a rate of about one hundred and sixty-seven miles each second, it is calculated that the first fearful outburst must have caused movement near the surface of the sun at a rate of at least three hundred miles each second. Probably the hydrogen cloud was borne upwards along with a vast mass of fragments flung out from the sun. We are here upon doubtful ground; but this stupendous power of eruption in the sun, and of driving matter out of and away from his surface, should not be forgotten.

Though such a sun-storm as that just described

is not often to be seen, yet there are at all times certain strange red prominences, or glowing flames, rising up here and there from the sun's "limb." Doubtless they rise also from other parts of the photosphere, though they are only visible to us when near enough to the edge to stand out beyond it.

As generally seen, these prominences have clear sharp outlines and are usually bright rose-red in color. They are described as sometimes wide and low, sometimes tall and slender; sometimes jagged, sometimes regular; sometimes keeping long the same shape, sometimes changing quickly in a few minutes. They are said to be like flames, like mountains, like the teeth of a saw, like icebergs, like floating cloudlets.

As to their height, from fifty to eighty thousand miles is nothing unusual. We must not speak of Mont Blanc or Mt. Everest here. Jupiter, placed bodily on the surface of the sun beside such a fire-mountain, would not far overtop it. The earth, Venus, Mars, and Mercury would lie like little toy-balls at its foot. And these are common-sized sun-flames. One has been measured which reached one hundred and sixty thousand miles in height, and even this has been exceeded.

The spectroscope shows these solar prominences or jets to be principally formed of glowing hydrogen gas.

Beyond the sierra or chromatosphere — that border of rippling crimson fiery billows round the edge of the sun, with tall red mountains rising out of it here and there—beyond these stretches the corona.

The corona, as seen from earth, is a bright far-reaching glory of light shining round the sun in a total eclipse. The moon then comes between the sun and the earth, her dark round body creeping over the face of the sun till the bright photosphere is completely covered. But the sierra and the red prominences stand out from behind the black moon, and the beautiful soft corona light stretches far beyond.

It was long doubted whether the corona really belonged to the sun or to the moon. No question now exists that it is a part of the sun.

Various descriptions of the corona have been given at different times as observed during different eclipses. It has been seen as a steady beamy white cloud behind the moon, showing no flickering. It has been seen marked with bright lines of light and seeming to move rapidly round and round. It has been seen silvery white, sending off long streams of brightness. It has been seen in the form of white light, with bluish rays running over it. It has been seen with entangled jets of light, like "a hank of thread in disorder." It has been seen silvery white again, with a faint tinge of greenish-violet about the outer edge. It has been seen from a high mountain-top as a mass of soft bright light, "through which shot out, as if from the circumference of the moon, straight massive silvery rays, seeming distinct and separate from each other, to a distance of two or three diameters of the lunar disc, the whole spectacle showing as upon a background of diffused rose-colored light."

The shape of the corona seems to change much at different times; and the outer edge is blurred and indistinct, fading imperceptibly away. How far it reaches, in a state of extreme thinness, no one can tell. Outlying portions, which the strongest sight cannot detect in the most powerful telescope, can be pictured by means of photography.

Different explanations of this crown of light have been suggested. Some suppose it to consist of an atmosphere, gaseous in kind, reaching outward from the sun's surface, growing ever more and more rare, after the fashion of our own atmosphere, only to an inconceivably greater extent. Some have conjectured it to be formed of countless revolving "meteorite particles," more after the fashion of Saturn's rings. But our knowledge of the true nature of the corona is yet in its infancy.

A few closing words as to the size and weight of the sun.

In diameter eight hundred and sixty thousand miles, and in bulk equal to over one million two hundred thousand earths, his weight is in proportion less.

Our earth is about four times as dense as the sun. If her size were increased to the sun's size, her density being the same as now, she would be very much heavier than the sun and would attract much more strongly.

Still, though the sun is of lighter materials than the earth, his immense size gives him weight equal to seven hundred and fifty times as much as all the planets put together.

The attraction on the surface of the sun is also very great—so great that we can hardly picture it to ourselves. If life exists there at all—supposing it possible that any kind of life can be in such a fiery atmosphere—it must be life very different from any known in this world.

A man who on earth weighs twelve or fourteen stone and walks lightly erect, would on the sun lie helplessly bound to the ground, crushed by his own overpowering weight.

It is said that a cannon-ball reposing on the sun, if lifted one inch and allowed to fall, would dash against the ground with a speed three times greater than that of our fastest express-trains.

For weight on earth is merely caused by the amount of force with which the earth draws downward a body towards herself—a force greater or less according to the density of that body. So weight on the sun would be immensely increased by his immensely greater power of attraction.

It is an interesting question how far the sun's attractive influence reaches effectually through space. The nearer a body is to the sun, the greater the attraction which he exercises over it.

At the distance of the planet Mercury a speed of some thirty miles each second is needful to overcome or balance it sufficiently for the planet to remain in his orbit. At the distance of the planet Neptune about four miles each second is enough. If a planet were journeying at four times the distance of Neptune, the speed would need to be not over two miles each second, lest the planet should break loose and wander away.

But even two miles a second is no mean speed—more than seven thousand miles an hour. If we come to speak of that which we on earth call rapid motion, we shall gain a clearer idea as to the extent of the sun's power.

Suppose a planet were travelling through space at the rate of one of our express-trains — sixty miles an hour. It has been calculated that, unless the sun's attraction were interfered with and overpowered by some nearer sun—which, by the way, would probably be the case — the said planet, though placed at a distance ten or twelve times as great as that of the far-off star Alpha Centauri, would still be forced by the sun's attraction to journey round him in a closed orbit. At such a speed it would not be free to wander off into the depths of space.

CHAPTER IV.

MORE ABOUT THE MOON.

"When I consider ... the moon ... which Thou hast ordained."—Psa. 8 : 3.

From a globe all fire, all energy, all action, we come to a globe silent, voiceless, changeless, lifeless.

So, at least, the moon seems to us. But it does not do to speak too confidently.

True, we can find no trace of an atmosphere in the moon. If there is any atmosphere at all, it must be so thin as to be less than that which we on earth count as actually none. We pump away the air from a glass inclosure, in an air-pump, and say the glass is empty.

Only it is not absolutely empty. There is always a very little air remaining; merely a few stray particles, perhaps, so little that fire would not burn and animals could not live in it. Very possibly air to that amount may still cling to the moon; but this is much the same as to say that no air at all is there. At all events, it is far too little to support life.

The question of the lunar atmosphere is full of interest. Many theories have been started to account for its absence; and most of these, sooner or later, have had to be given up.

Of late a singular thought has come to the front — that the moon very likely had once an atmosphere, like other worlds, but that in course of time she lost it as a matter of necessity.

Why should she?

Because she was too small to keep it.

Something may certainly be said in support of this view. An atmosphere is made of gases; and the tiny particles of a gas are in swift and ceaseless motion, fighting to get as far off one from another as possible. Such motion would carry the particles of every gas away from every planet and sun, except for the restraint of gravitation. As our earth by the force of attraction holds down her atmosphere, so do other celestial bodies.

But if a world be very small, then its attraction is weak, too weak, it may be, for the control of those swift gas particles. In such a case the atmosphere might be expected, in the course of years, slowly to break loose, particle by particle, melting away into distant space.

This *may* have been the case with our moon.

She is so little, and gravitation on her surface must be so feeble, that she might have failed to hold in the restless atmosphere which, perhaps, once clothed her.

For the moon is very much smaller than the earth. Her diameter is about two-sevenths of the earth's diameter; her entire surface is about two-twenty-sevenths of the earth's surface; her size is about two-ninety-ninths of the earth's size; and her whole weight is about one-eightieth of the earth's weight.

Attraction or gravitation on the surface of the moon is very different from what it is on the earth. Her much smaller bulk greatly lessens her power of attraction. While a man from earth would on the surface of the sun—supposing he could exist there at all—lie helpless, motionless, and crushed by his own weight, he would on the moon find himself astonishingly light and active. A leap over a tall house would be nothing to him.

The moon, unlike the sun, has no light or heat of her own to give out. She shines merely by reflected light. Rays of sunlight falling upon her rebound thence and find their way earthward.

This giving of reflected light is not a matter

all on one side. We yield to the moon a great deal more than she yields to us. Full earth, seen from the moon, covers a space thirteen times as large as full moon seen from earth.

Perhaps you may have noticed soon after new moon, when a delicate crescent of silver light shows in the sky, that within the said crescent seems to lie the body of a round dark moon, only not perfectly dark. It shows a faint glimmer.

That glimmer is called earth-shine. The bright crescent shines with reflected sunlight. The dim portion shines with reflected earth-light.

What a journey those rays have had! First leaving the sun, flashing through ninety-two millions of miles to earth, rebounding from earth and flashing over two hundred and forty thousand miles to the dark shaded part of the moon, then once more rebounding and coming back, much wasted and enfeebled, across the same two hundred and forty thousand miles, to shine dimly in your eyes and mine.

Now about the *phases* of the moon, that is, her changes from " new " to " full " and back again to " new."

If the moon were a star-like body, shining by

her own light, she would always appear to be round. But as she shines by reflected sunlight, and as part of her bright side is often turned away from us, the size and shape of the bright part seem to vary.

For, of course, only that half of the moon which is turned directly towards the sun is bright. The other half turned away is dark, and can give out no light at all, unless it has a little earth-shine to reflect.

As the moon travels round the earth she changes gradually from new to full moon and then back to new again.

"New moon" is when the moon in her orbit comes between the sun and the earth. The half of her upon which the sun shines is turned away from us and only her dark side is towards us. So at new moon she is quite invisible. It is at new moon that an eclipse of the sun takes place, whenever the moon's orbit carries her in a line precisely between sun and earth.

Passing onwards round the earth, the moon, as we get a little glimpse of her shining side, first shows a slender sickle of light, which widens more and more till she reaches her first quarter. She is then neither between earth and sun nor outside the earth away from the sun, but just at one side of us, passing over the earth's own orbit. Still, as before, half her body is lighted up by the sun. By this time *half* the bright part and *half* the dark part are turned towards us; so that, seeing the bright quarter, we name it the "first quarter."

On and on round us moves the moon, showing more light at every step. Now she passes quite outside the earth's orbit, away from the sun. Not the slightest chance here of an eclipse of the sun, though an eclipse of the moon herself is quite possible. But more of that presently. As she reaches a point in a line with earth and sun—only generally a little higher or lower than the plane of the earth's orbit—her round bright face, shining in the sun's rays, is turned exactly towards us. Then we have "full moon."

Still she goes on. Once more her light narrows and wanes, as part of her bright half turns away. Again, at the "last quarter" as at the first, she occupies a "sideways" position, turning towards us half her bright side and half her dark side. Then she journeys on with lessening rim of light till she vanishes, and once more we have the dark invisible "new moon."

You may easily picture this for yourself. Let a candle be upon a table in a room otherwise dark. Stand near the table with a ball in your hand, which you must hold out at arm's length. Turn slowly round upon your feet, keeping your arm and hand motionless, and let the ball thus travel round you.

Your own motion here is a matter of no importance. It is needful merely that you may keep your face towards the ball, and also that the position and slope of the ball may not vary.

Begin by stretching out your arm straight towards the candle. Now the light is shining full upon the ball, lighting up one half of it. But of the bright half you see nothing. The dark side only is turned towards you. That is "new moon." If you put the ball exactly between your eyes and the candle it eclipses or hides the candle-flame, but by holding it an inch or two higher or lower you avoid an eclipse.

Now move your arm slowly round to the left, moving yourself round with it so as to watch the ball intently. A rim of light will begin to show, gradually increasing, till of the half-ball which you can see, one quarter is light and one quarter is dark. In the case of the real moon the light quarter only is visible and the dark is hidden, though in your ball you will see both. This is the "first quarter."

Go on turning slowly round, watching the light space gradually widen. Now your arm is extended in a direction away from the candle. If the ball is in an exact line with the candle and your head, you will produce an "eclipse of the moon" by allowing your shadow to fall upon it. But by raising it a little higher you avoid this eclipse also. The candle-light still, as always, falls upon the ball, lighting up one half; and now the whole of that half is turned towards you and the whole of the dark side is turned away. This is "full moon."

Move onward again, spinning very slowly round with extended arm, and you will have diminishing light once more, till the ball passes through "last quarter" and again becomes new moon.

An eclipse of the sun has been already described. An eclipse of the moon is an equally simple matter.

An eclipse of the sun is caused by the dark solid body of the moon passing just between earth and sun, hiding the sun from us and casting its shadow upon the earth.

An eclipse of the moon is also caused by a shadow—the shadow of our own earth—falling upon the moon.

Here again, if the plane of the moon's orbit were the same as ours, eclipses of the moon would be very common. As it is her orbit carries her often just a little too high or too low to be eclipsed; and it is only now and then, at regular intervals, that she passes through the shadow of the earth.

If a large solid ball is hung up in the air, with bright sunlight shining on it, the sunlight will cast a *cone of shadow* behind the ball. It will throw in a direction just away from the sun a long round shadow the same size as the ball at first, but tapering gradually off to a point. If the ball is near the ground, a round shadow will rest there almost as large as the ball. The higher the ball is placed, the smaller will be the round shadow, till at length, if the ball be taken far enough upwards, the shadow will not reach the ground at all.

Our earth and all the planets cast just such tapering cones of dark shadow behind them into space. The cone always lies in a direction exactly away from the sun.

It is when the moon comes into this shadow that an "eclipse of the moon" takes place. Sometimes she only dips half-way into it or just grazes along the edge of it, and that is called a "partial eclipse." Sometimes she goes in altogether, straight through the midst of the shadow, so that the whole of her bright face for a short time grows quite dark. Then we have a "total lunar eclipse."

CHAPTER V.

YET MORE ABOUT THE MOON.

"Our God is in the heavens; He hath done whatsoever He hath pleased." Psa. 115:3.

THERE are two ways of thinking about the moon. One way is to consider her as merely the earth's attendant satellite. The other way is to consider her as our sister-planet, travelling with us round the central sun.

The first is the more common view; but the second is just as true as the first.

For the sun does actually pull the moon towards himself with a very much stronger pulling than that of the earth. The attraction of the sun for the moon is more than double the attraction of the earth for the moon. If it were not that he pulls the earth quite as hard as he pulls the moon, he would soon overpower the earth's attraction and drag the moon away from us altogether.

People are often puzzled about the orbit or pathway of the moon through the heavens. For in one sense they have to think of her as travelling round and round in a fixed orbit, with the earth in the centre. In another sense they have to think of her as always journeying onward with the earth in her journey round the sun, and thus never returning to the same point.

There are two ways of meeting this difficulty.

First of all, remember that the one movement does not interfere with the other. Just as in the case of the earth travelling round the sun and also travelling onward with him through space, just as in the case of a boy walking round and round a mast and also being borne onward by the moving vessel, so it is here. The two movements are quite separate and independent of each other.

As regards the earth alone, the moon journeys round and round perpetually, not in a circle, but in a pathway which comes near being an ellipse.

As regards the actual *line* which the moon's movements may be supposed to draw in space, it has nothing elliptical about it, since no one point of it is ever reached a second time by the moon.

But according to this last view of the question, nobody ever can or will walk in a circle or an oval. Take a walk round your grass-plot, measuring your distance carefully at all points from the centre. Is that a circle? All the while you moved the surface of the earth was rushing along and bearing you with it, and the whole earth was hurrying round the sun, and was being also carried by him in a third direction. Whatever point in space you occupied when you started, you can *never fill that particular part of space again.* The two ends of your so-called circle can never be joined.

But then you may come back to the same point *on the grass* as that from which you started. And this is all that really signifies. Practically you have walked in a circle. Though not a circle as regards space generally, it is a circle as regards the earth.

So also the moon comes back to the same point *in her orbit round the earth.* Letting alone the question of space, and considering only the earth, the moon has—roughly speaking—journeyed in an ellipse.

You may, however, look at this matter in quite another light.

Forget about the moon being the earth's satellite, and think of earth and moon as two sister-planets going round the sun in company.

The earth, it is true, attracts the moon. So also the moon attracts the earth, though the far greater weight of the earth makes her attraction to be far greater. If earth and moon were of the same size they would pull each other with equal force.

But though the pull of the earth upon the moon is strong, the pull of the sun upon the moon is more than twice as strong. And greatly as the earth influences the moon, yet the actual centre of the moon's orbit is the sun and not the earth. Just as the earth travels round the sun, so also the moon travels round the sun.

The earth travels steadily in her path, being only a little swayed and disturbed by the attraction of the moon. The moon, on the contrary, while travelling in her orbit is very much swayed and disturbed indeed by the earth's attraction. In fact, instead of being able to journey straight onwards like the earth, her orbit is made up of a succession of delicate curves or scallops passing alternately backwards and forwards over the orbit of the earth. Now she is behind the earth, now in front of the earth; now between earth and sun, now outside the earth away from the sun. The order of positions is not as here given, but each is occupied by her in turn. Sometimes she moves quickly, sometimes she moves slowly, just according to whether the earth is pulling her on or holding her back.

Two hundred and forty thousand miles sounds a good deal. That is the medium distance of earth from moon. But it is, after all, a mere nothing compared with the ninety-two or ninety-three millions of miles which separate the sun from the earth and moon.

If we made a small model, with the sun in the centre and the earth and moon travelling a few inches off, only one slender piece of wire would be needed to represent the path of earth and moon together. For not only would the earth and the moon be so small as to be quite invisible, but the whole of the moon's orbit would have disappeared into the thickness of the single wire.

This question of the moon's motions is in its nature intricate and in its details quite beyond the grasp of any beginner in astronomy.

But so much at least may be understood, that though the earth's attraction powerfully affects the moon, and causes in her motions *perturbations* such as have been already spoken about as taking place among the planets, yet in reality the great controlling power over the moon is the attraction of the sun.

The tides of the ocean are chiefly brought about by the moon's attraction. The sun has something to do with the matter, but the moon is the chief agent.

This action of the moon is best seen in the southern hemisphere, where there is less land. As the moon travels slowly round the earth her attraction draws up the yielding waters of the ocean in a vast wave, which travels slowly along with her. The same pulling which thus lifts a wave on the side of the earth towards the moon also pulls the earth gently *away from* the water on the opposite side, and causes a second wave there. The parts of the ocean between these two huge waves are depressed, or lower in level.

These two waves on opposite sides of the earth sweep steadily onward, following the moon's movements—not real but seeming movements, caused by the turning of the earth upon her axis.

Once in every twenty-four hours these wide waves sweep round the whole earth in the southern ocean. They cannot do the same in the north, on account of the large continents, but offshoots from the south waves travel northward, bringing high tide into every sea and ocean inlet.

If there were only one wave, there would be only one tide in each twenty-four hours. As there are two waves, there are two tides, one twelve hours after the other. In the space between these two high tides we have low tide. ·

Twice every month we have the greatest rise and fall of the tides. These are called "spring tides." Twice every month we have the least rise and fall. These are called "neap tides."

When the moon is between us and the sun, or when she is "new moon," there are spring tides; for the pull or attraction of sun and moon upon the ocean act exactly together. It is the same at full moon, when once more the moon is in a straight line with earth and sun.

But at the first and last quarters, when the moon has her *sideways* position, and when the sun pulls in one direction and the moon pulls in another, each undoes a little of the other's work. Then we only have neap tides, for the wave raised is smaller and the water does not flow so high upon our shores.

In speaking of the surface of the moon we are able only to speak about one side. The other is entirely hidden from us. This is caused by the curious fact that the moon turns on her axis and travels round the earth in exactly the same length of time.

One half of the moon is thus always turned towards us, though of that half we can only see so much as is receiving the light of the sun. But the half turned in our direction is always the same half.

One part of the moon—not quite so much as half, though always the same portion—is turned away from us. A small border on each side of that part becomes now and then visible to us, owing to certain movements of the earth and the moon.

What sort of a landscape may lie in the unknown district it is idle to imagine. Many guesses have been made. Some have supposed it possible that air *might* be found there, that water *might* exist there, that something like earthly animals *might* live there.

These, however, are now notions of the past, no longer to be entertained. Judging from mere earthly experience, nothing seems more unlikely than that air and water should be banished from one half of a world and collected together in the space remaining. Moreover, as already explained, we know how extremely natural a thing it is that the moon should have lost her atmosphere through the weakness of her attractive power;

and this would affect equally all sides of the globe. If any air at all exists there, it must be in density less than the two-thousandth part of what we breathe.

It seems almost equally clear that water must be entirely absent. No signs of water-action are visible in the craggy mountains and deep craters.

If there at all, it could only be in the form of ice, since the absence of an atmosphere keeps the surface of the moon, even during the glare of its long day, below zero.

The craters which honeycomb the surface of the moon are various in size. Many of the larger ones are from fifty to a hundred miles in diameter. These huge craters, or, as we may call them, deep circular plains, are surrounded by mighty mountain ramparts rising to the height of thousands of feet. Usually they have in their centre a sugar-loaf or cone-shaped mountain, or even two or more such mountains, somewhat lower in height than the surrounding range.

The sunset-lights upon certain of these distant mountain-peaks were first watched by Galileo through his telescope, and have since been seen by many an observer—intense brightness contrasting with intense blackness of shadow.

In addition to her great craters the moon seems to be thickly covered with little ones, many being as small as can be seen at all through a telescope.

Whether these are all volcano-craters remains to be discovered. It is not supposed that any of them are now active. From time to time signs of faint changes on the moon's surface have been noticed, which it was thought might be owing to volcanic outbursts. Such an outburst as the worst eruptions of Mt. Vesuvius would be invisible at this distance. But the said changes may be quite as well accounted for by the startling fortnightly variations of climate which the moon has to endure. The general belief now inclines to the idea that the moon-volcanoes are extinct, though no doubt there was in the past great volcanic activity there.

A description has been given earlier of the rain of meteorites constantly falling to our earth, and only prevented by the atmosphere from becoming serious.

The moon has no such protecting atmosphere, and the amount of cannonading which she has to endure must be by no means small. Perhaps in past times, when her slowly-cooling crust was yet soft, these celestial missiles showering upon her may have occasionally made deep round holes in her surface.

This is another guess, which time may prove to be true. Guesses at possible explanations of mysteries do no harm so long as we do not accept them for truth without ample reason.

Beside the craters and their surrounding barriers there are ranges of mountains on the moon, and flat plains which were once named "seas" before it was found that water did not exist there. Astronomers also see bright ridges, or lines, or cracks of light, hard to explain.

One of the chief craters is called "Ptolemy," and in size it is roughly calculated to be no less than one hundred and fourteen miles across. Another, "Copernicus," is about fifty-six miles, and another, "Tycho," about fifty-four miles. The central cone-mountain of Tycho is five thousand feet high. The crater of "Schickard" is supposed to be as much as one hundred and thirty-three miles in diameter.

The so-called "seas" are those large dark spots to be seen on the moon's surface in the shape of "eyes, nose, and mouth," or of the famous old man with his bundle of sticks. The brighter parts are the more mountainous parts.

The chief ranges of lunar mountains have been named by astronomers after mountains on earth, such as the Apennines, the Alps, the Caucasian range, the Carpathian, and the Altai Mountains.

NOTE.—With reference to the question of tides, slightly touched upon in this chapter, a few words of additional explanation have been kindly sent to me by a naval officer, so clear and concise that I insert them as they stand:

"When the tidal wave is advancing it causes a current to run in the direction of its advance. This current is called the 'flood tide.'

"When the highest part of the tidal wave has reached any place it is said to be 'high water' at that place.

"Very soon after high water the receding of the tidal wave causes the current to run in the opposite direction to the flood, and to continue running till the lowest point of the tidal wave has been reached. This receding current is called the "ebb tide," and its lowest point is called 'low water.'

"Both at high and low water there is a brief period when the tide neither ebbs nor flows. This period is called 'slack water.'"

CHAPTER VI.

MERCURY, VENUS, AND MARS.

"Thou hast made the heaven and the earth by Thy great power and stretched-out arm, and there is nothing too hard for Thee."—JER. 32:17.

ONCE again we have to journey through the high-roads of the Solar System, paying a brief visit to each in turn of our seven chief brother-and-sister planets and learning a few more leading facts about them. Having gone the same way before, it will not now seem quite so far.

Busy hurrying Mercury! we must meet him first in his wild rush through space. If he were to slacken speed for a single instant, he would begin to fall with fearful rapidity towards the sun. And if Mercury were to drop into one of those huge black chasms of rent furnace-flame on the sun's surface, there would be a speedy end to his life as a planet.

Mercury's year is about one quarter the length of our year; and his day is believed by Schiaparelli, from certain observations made, to be identical in length with the planet's year. This, however, is as yet by no means certain. If true, it would mean that Mercury revolves round the sun, presenting to him always one side only, and never the other, just as our moon revolves round the earth.

Whether Mercury's axis slopes like earth's axis we do not yet know. More can be said about his orbit. The earth's orbit, as before explained, is not a circle, but an ellipse or oval. Mercury's orbit is an ellipse also, and a much longer—or, as it is called, a more eccentric—ellipse. The earth is three millions of miles nearer to the sun at one time of the year than six months before or after. Mercury is no less than fifteen millions of miles nearer at one time than another, which must make a marked difference in the amount of heat received.

Even when the distance is greatest, the sun as seen from Mercury looks more than four times as large as the sun we see.

What a blazing splendor of light! It is not easy to imagine human beings living there in such heat and glare, and with either no changes of season at all or such very short seasons rapidly following one another. Mercury may, not impossibly, abound with living creatures of some sort;

but they would have to be an altogether different kind of creatures from any seen on earth. True, a very considerable power of adaptation is found here in many animals, especially in man; so that great varieties of climate, from extreme heat to extreme cold, can be endured. It is not easy to say how far such power of adaptation to circumstances may reach. Still, the extremest heat ever felt on earth is as nothing beside that of Mercury; and the common-sense view of things is very much opposed to the possibility of any life on Mercury identical in kind with any life known on earth.

The suggestion has been made that a moist dense atmosphere might possibly help to ward off the overpowering heat and to make the planet more habitable. Experience, however, argues against this theory. Our earthly atmosphere acts in a precisely contrary fashion, serving as a storehouse for heat, and rendering our planet much warmer than if no air existed.

There is, after all, no need to conclude hastily that all these worlds in our neighborhood are, or must be, inhabited at this present time. Why, indeed, should they? With perfect ease each world *could* have been adapted by divine power to its inhabitants, or the inhabitants to each world. But, quite conceivably, they may be destined for some entirely different purpose, lying outside the range of our knowledge and imagination. Or they may be still in process of preparation for future inhabitants, as, during countless ages, our earth was in preparation for the little span of man's life upon her surface. We need not thrust our earthly notions of haste into the thought of the vast cycles of a universe formed in the infinite leisure of the divine counsels.

If, in addition to a possible atmosphere, Mercury were surrounded habitually, as some have conjectured, by an envelope of heavy clouds, this might no doubt, to some extent, modify the desperate heat.

The small size of Mercury makes attraction on his surface much less than on earth. A lump of iron weighing on earth one pound would weigh on Mercury only about seven ounces, or less than half as much. So a man would be a very light leaper indeed there, and an elephant might be quite a frolicsome animal.

If there should be star-gazers in Mercury, the earth and Venus must both be beautiful to look upon. Each of the two would shine far more brightly than Jupiter as seen at his best from earth.

Like Mercury, Venus, the next planet, has an orbit lying inside our orbit. Mercury and Venus are always nearer to the sun than we are.

And if Mercury and Venus travelled round the sun in orbits the planes of which were exactly the same as the plane of the earth's orbit, we should very often see them creeping over the surface of the sun.

Not that they really "creep over" it; only, as they journey between the sun and us, we can see them pass like little black dots across the sun's disc. This is the same thing as when the moon passes across the sun's disc and eclipses it. But Mercury and Venus are too far away from us to cause any eclipse of the sun's light.

These crossings of the sun's face, or "transits," as they are called, have been important matters. The transit of Venus especially was once eagerly looked for by astronomers, since by close observations of Venus' movements and positions the distance of the sun could at that time be better calculated than in any other way. Other methods are now coming into vogue.

The transits of Venus are rare. Two come near together, separated by only eight years, and then for more than one hundred years the little dark body of Venus is never seen from earth to glide over the sun's photosphere. There was a transit of Venus in the year 1761 and another in the year 1769. There was a transit of Venus in 1874, and there was also another in 1882. At the last transit it was found that the sun was only about ninety-one or ninety-two millions of miles away, while later calculations of a different kind have fixed the amount as close upon ninety-three millions.

The reason why these transits happen so seldom is that the orbits of Mercury and Venus lie in rather a different plane or level from the earth's orbit. So, like the moon, though often passing between us and the sun, they generally go just a little higher or just a little lower than his bright face.

Mercury and Venus show phases like the moon, although they do not circle round the earth as the moon does. These "phases," or changes of shape, are probably never visible except through a telescope.

It will be easier to think about the phases of

Venus alone than to consider both together. Her orbit lies within the earth's orbit, and the earth and Venus travel round the sun—as do all the planets—in the same direction. But as Venus' pathway is shorter than ours, and as her speed is greater, she is much the quicker about her yearly journey, and she overtakes us again and again at different points of our orbit in turn.

At one time she comes between us and the sun. That is her nearest position to us, and she is then only about twenty-five millions of miles distant. A beautiful sight she would be, but unfortunately her bright side is entirely turned away and only her dark side is turned towards us. So then she is "new Venus," and is invisible.

At another time she is completely beyond the sun and at her farthest position away from us. Her shining is quite lost in the sun's rays coming between. And though we get a good view of her as "full Venus" at a little to one side or the other, yet so great is her distance—as much as one hundred and fifty-seven millions of miles—that her size and brightness are very much lessened.

Between these two nearest and farthest points she occupies two middle distances, one on each side of the sun. Then, like the moon at her "quarters," she turns to us only half of her bright side. But this is the best view of Venus that we have, as a brilliant untwinkling star-like form— the Evening Star of ancients and of poets.

Between these four leading positions Venus is always travelling gradually from one to another— always either waxing or waning in size and in brightness.

Mercury passes through the same seeming changes.

If Venus owned inhabitants they would have a glorious view of the earth with her moon. At the time when the two planets are nearest together, and when she is only "new Venus" to us, a dark and invisible body, the earth is "full earth" to Venus. The very best sight we ever have of Venus cannot come near that sight. But if Venus should be so often covered with heavy clouds as some have thought possible, this must greatly interfere with any habits of star-gazing.

Venus and the earth have often been called twin-sister planets. There are many points of likeness between them. In size at least they differ little. Venus, as well as Mercury, is believed by Schiaparelli to have a day equal in length to the planet's year, one face only of the planet being thus turned always downward. M. Trouvelot, on the contrary, as a result of other observations, reverts to the old idea that Venus has a day like in length to that of her twin-sister, Earth—about twenty-four hours long. It is impossible as yet to know which will prove to be in the right.

M. Trouvelot also believes that certain bright spots occasionally seen on Venus are very lofty mountains showing their peaks above the thick atmosphere. This again is a reversion to, or at least an additional proof of, an idea many years old, that mountains on Venus tower to three or four times the height of our stateliest earthly summits.

There is a good deal of uncertainty about the climate of Venus. The heat there must greatly surpass heat ever felt on earth — the sun being about double the apparent size of our sun and pouring out nearly double the amount of light and heat that we receive in our tropics in the hottest day ever known. Human beings, accustomed to earthly heat, could scarcely exist or keep their eyesight under such a blaze and glare as this must imply.

It it believed also that the axis of Venus, instead of being slanted only as much as the earth's axis, is tilted much more. Even if the tilting is less than some have supposed, it is probably very considerable. If Venus really does "lie over" in such a manner, certain startling changes of climate, unpleasant changes according to our ideas, would take place.

Like earth, Venus would have her two arctic regions, where a burning summer's day would succeed a bitter winter's night, each half a year in length. She would have also her tropical region, only in that region intense cold would alternate with intense heat—brief seasons of each in turn. And between the tropics and the arctic regions would lie wide belts, by turns entirely tropical and entirely arctic. The rapidity and severity of these changes, following one another in a year about as long as eight of our months, would seem to be too much for any human frame to endure.

But it all rests upon an *if*. And we may be quite sure that *if* there are any living creatures in Venus their frames are well suited to the climate of their world.

Though Mars is one of the inner group of four small planets, divided by the zone of asteroids from the outer group of four great planets, yet he belongs to the outside set of Superior Planets. His orbit surrounds ours, being at all points farther off from the sun.

Very slight "phases" have been seen in Mars. He turns to us from time to time just enough of his dark side to prove that he *has* a dark side and that he does not shine like a star by his own light. But the phases are by no means marked as with Venus.

It was long believed that Mars possessed no moons. Two very small ones have, however, been lately found circling round him. They have been named Deimos and Phobos, after the "sons of Mars" in Greek mythology. Deimos travels round Mars in thirty-nine hours, while Phobos performs the same journey in the astonishingly short period of seven hours and a half!

Mars is not only much smaller than the earth, but a good deal less dense in his "make." His material is only about three-quarters as heavy as an equal amount of the earth's material. A very heavy man on earth would be a most light and active individual on Mars. Gold taken from earth to Mars would weigh there no more than tin weighs upon earth.

There are good reasons for believing that Mars has an atmosphere. Indeed it has been conjectured that Mars may possibly be somewhere about one of the smallest-sized worlds able permanently to retain their atmospheres by force of gravitation.

The surface of Mars has been more studied, because it *can* be more studied, than that of any other planet in our system. It is well placed for observation, being outside the earth instead of between us and the sun, and it is also comparatively near. True, at his very nearest, Mars does not draw closer than thirty-five millions of miles, and one has to be very cautious in making statements about a world thirty-five millions of miles away. But this is not much compared with the distance of Jupiter.

And the surface of Mars presents features well worthy of study, albeit perplexing. Here and there, it is true, cloud-forms seem to sweep over the landscape, blotting out what lies below. Still, though such appearances come and go, they do go as well as come! Mars is not, like Jupiter and Saturn, permanently wrapped in such masses of vapor that one cannot tell whether any solid body at all lies within.

The reddish color of Mars is well known. This does not vanish in the telescope, but it is found that parts only have the reddish hue, while other parts are dark and greenish. These markings have been so closely examined that more is known about the geography of Mars than of any other world outside of our own. It is supposed that they *may* be continents and oceans, and names have been given to them, such as Dawes Continent, Herschel Continent, Airy Sea, Huggins' Inlet, and so on.

Land and water—if these markings really are such—seem to be very differently distributed on Mars from what they are on earth. Here we have about three times as much water as land, and to get from one continent to another without crossing the sea is in some cases impossible.

But a traveller there might go most conveniently to and fro, hither and thither, to all parts of his world, either on land or on water, without any change. If he preferred water, he would never need to set foot on land, and if he preferred land, he would never need to enter a boat. The two are so curiously mingled together, narrow necks of land running side by side with long, narrow sea-inlets, that Atlantic and Pacific Oceans are unknown.

These landscape features of the planets seem to be not absolutely changeless, like the continents and oceans of earth, and of course any alterations seen by us from earth, in the course of a few years, must imply vast changes of outline there. It may be that Mars is now in a state which earth passed through many ages ago, a state of rapid geological change, rising and sinking of continents and shifting of oceans. But it is at least quite as probable that the changes are in the main atmospheric, and that the continents and canals really alter very little.

If clouds do actually float in the atmosphere of Mars, this means the existence of water in Mars. Doubtless rain sometimes falls there.

Two singular white spots are to be seen at the north and south poles which are most probably polar ice and snow. Anybody looking at our earth in like manner from a distance, might perceive two such white snow-caps. These caps are seen to vary with the seasons. When the north pole of Mars is turned towards the sun the white spot there grows smaller; and at the same time, the south pole of Mars being turned away from the sun, the white spot there grows larger. Again, when the south pole is towards the sun, and the north pole away from the sun, the white spot at the south is seen to be the smallest and the white spot at the north is seen to be the largest. This is exactly what takes place in the summers and winters of our north and south poles.

It is rather curious that Mars' caps should be much smaller than ours, since Mars is farther from the sun. But Mars is thought to contain proportionately less water than our earth does; and a drier atmosphere would at once account for a lessened amount of snow.

CHAPTER VII.

JUPITER.

"Whatsoever the Lord pleased, that did He in heaven and in earth."—PSA. 135:6.

PASSING at one leap over the belt of tiny asteroids, about which we know little beyond their general movements and the size and weight of a few among them, we reach at once the giant planet Jupiter—mighty Jupiter, hurrying ever onward with a speed not indeed equal to that of Mercury or of our earth, yet eighty times as rapid as the speed of a cannon-ball!

Think of a huge body, equal in bulk to twelve hundred earths and in weight to over three hundred earths, rushing ceaselessly through space at the rate of seven hundred thousand miles a day!

Jupiter's shape is greatly flattened at the poles. He spins rapidly on his axis once in nearly ten hours, and has therefore a five hours' day and a five hours' night. As the slope of his axis is exceedingly slight, he can boast little or no changes of season. The climate near the poles has never much of the sun's heat. In fact all the year round the sun must shine upon Jupiter much as he shines on the earth at the equinoxes.

But the amount of light and heat received by Jupiter from the sun is only about one twenty-fifth part of that which we receive on earth; and the sun as seen from Jupiter can have but a small round surface not even one-quarter the diameter of the sun we see in the sky.

When looked at with magnifying power the bright star-like Jupiter grows into a broad, softly-shining disc or plate, with flattened top and bottom, and four tiny bright moons close at hand. Sometimes one moon is on one side and three are on the other; sometimes two are on one side and two on the other; sometimes one or more are either hidden behind Jupiter or passing in front of him. Jupiter has also curious markings on his surface, visible through a telescope. These markings often undergo changes, for Jupiter is no chill, fixed, dead world such as the moon seems to be.

There are dark belts and bright belts, usually running in a line with the equator, from east to west. Across the regions of the equator lies commonly a band of pearly white, with a dark band on either side of " coppery, ruddy, or even purplish " hue. Light and dark belts follow one after another up to the north pole and down to the south pole.

When we talk of "north and south poles" in the other planets, we merely mean those poles which point towards those portions of the starry heavens which we have chosen to call "northern" and "southern."

You know that all the chief planets travel round the sun in very nearly the same *plane* or flat surface that we do ourselves. That plane is called the "plane of the ecliptic."

Suppose that you had an enormous sheet of cardboard, and that in the middle of this cardboard the sun were fixed, half his body being above and half below. At a little distance, fixed in like manner in the card, would be the small body of the earth, half above and half below, her axis being in a slanting position.

The piece of cardboard represents what is called in the heavens the *plane of the ecliptic*—an imaginary flat surface cutting exactly through the middle of the sun and of the earth.

If the planets all travelled in the same precise plane, they would all be fixed in the cardboard just like the earth, half the body of each above and half below. As they do not so travel, some would have to be placed a little higher, some a little lower, according to what part of their orbits they were on.

This supposed cardboard "plane of the ecliptic" would divide the heavens into two halves. One half, containing the constellations of the Great Bear, the Little Bear, Cepheus, Draco, and others, would be called the Northern Heavens. One end of the earth's axis, pointing just now nearly to the Polar Star, we name the North Pole; and all poles of planets pointing towards this northern half of the heavens are in like manner named by us their north poles.

With regard to west and east, lay in imagination upon this cardboard plane a watch, with its face upward, remembering that all the planets and nearly all the moons of the Solar System are said both to spin on their axes and to travel in their orbits round the sun *from west to east*. Note how the hands of your watch would move in such a position. The "west to east" motions of planets and moons would be in exactly the opposite direction from the motions of the watch-hands.

To return to Jupiter. It is believed that these bands of color are owing to a heavy dense atmosphere loaded with vast masses of cloudy vapor.

By the "size" of Jupiter we really mean the size of this outside envelope of clouds. How large the solid body within may be, or whether there is any such solid body at all, we do not know. The extreme lightness of Jupiter as compared with his great size has caused strong doubts on this head.

The white belts are supposed to be the outer side of cloud-masses shining in the sunlight. Travellers in the Alps have seen such cloud-masses spreading over the whole country beneath their feet, white as driven snow, and shining in the sunbeams which they were hiding from villages below, or looking like masses of cotton-wool, from which the mountain-peaks rose sharply here and there.

The dark spaces between seem to be rifts or breaks in the clouds. Whether, when we look at those dark spaces, we are looking at the body of Jupiter, or only at lower layers of clouds, is not known. But sometimes blacker spots show upon the dark cloud-belts, and this seems rather as if they were only lower layers of clouds, the black spots giving us peeps down into still lower and deeper layers, or else perhaps to the planet itself.

These appearances remind one strongly of the sun-spots, each with its penumbra, umbra, and nucleus.

Occasionally bright white spots show instead of dark ones. It is thought that they may be caused by a violent upward rush of dense clouds of white vapor. The white spots again recall the sun and his *faculæ*.

Jupiter's bands are not fixed. Great changes go on constantly among them. Sometimes a white band will turn dark-colored or a dark band will turn white. Sometimes few and sometimes many belts are to be seen. Sometimes a dark belt will lie slanting across the others, nearly from north to south. Once, in a single hour, an entirely new belt was seen to come into shape. Another time two whole belts vanished in one day. The bands in which such rapid movements are seen are often thousands of miles in breadth.

Sometimes these wide zones of clouds will remain for weeks the same. At another time a break or rift in them will be seen to journey swiftly over the surface of the planet.

The winds on earth are often destructive. A hurricane moving at the rate of ninety miles an hour will carry away whole buildings and level entire plantations. Such hurricanes rarely if ever last more than a few hours.

But winds in Jupiter, judging from the move-

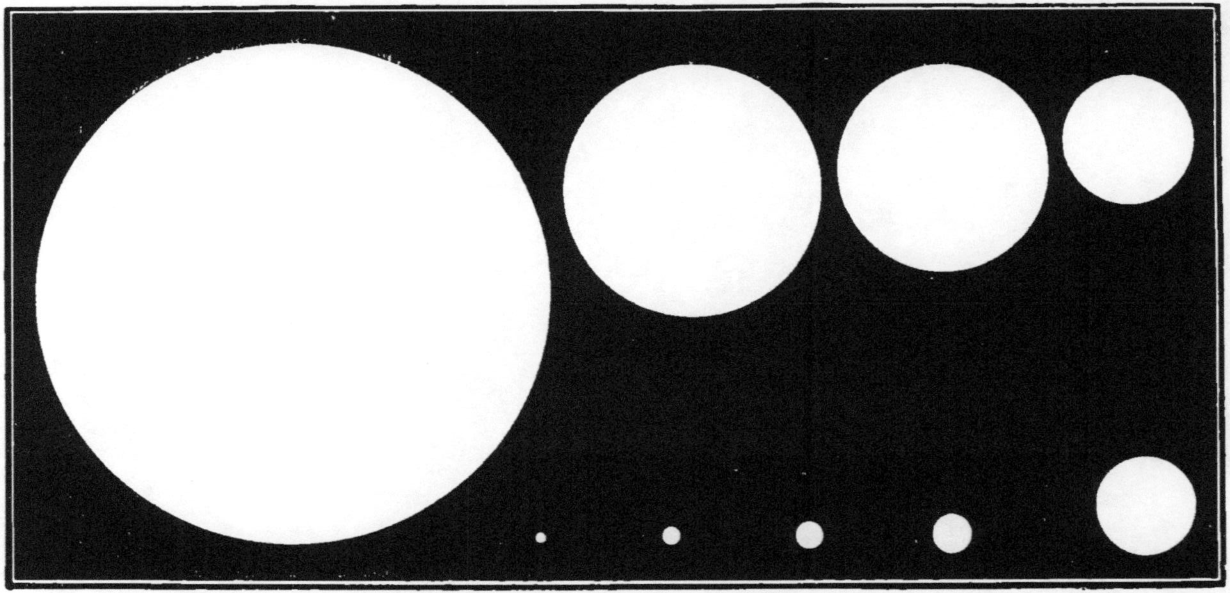

APPARENT MAGNITUDE OF THE SUN, AS SEEN FROM THE DIFFERENT PLANETS.

ments of the clouds, often travel at the rate of one hundred and fifty miles an hour, and that not for hours only, but for many weeks together. What manner of living beings could stand such weather may well be questioned.

Another difficulty which arises is as to the cause of these portentous disturbances on Jupiter. Our earthly storms are brought about by the heat of the sun acting on our atmosphere. But the sun-heat which reaches Jupiter seems far from enough to raise such vast clouds of vapor and to bring about such prolonged and tremendous hurricanes.

What if there is another cause? What if Jupiter is *not* a cooled body like our earth, but a liquid seething, bubbling mass of fiery heat—just as we believe our earth was once upon a time, in long past ages, before her outside crust became cold enough for men and animals to live thereon?

Then indeed we could understand how, instead of oceans lying on his surface, all the water of Jupiter would be driven aloft to hang in masses of steam or be condensed into vast cloud-layers. *Then* we could understand why a perpetual stir of rushing winds should disturb the planet's atmosphere.

In that case would Jupiter be a planet at all?

Certainly—in the sense of obeying the sun's

control. Our earth was once, we believe, a globe of melted matter, glowing with heat, and, farther back still, possibly, a globe of gas. Some people are very positive about these past changes; but it is wise not to be over-positive where we cannot know to a certainty what has taken place. However, Jupiter *may* have cooled down only to the liquid state, and if he goes on cooling he may by-and-by gain a solid crust like the earth. Times of long and slow preparation are often needed alike by individuals, by nations, and by worlds. If Jupiter is ever to be an inhabitable planet, that goal must lie still a long way ahead.

A few words as to the moons ever circling round and round the body of Jupiter.

Slight mention has been made earlier of a fifth moon, recently detected by Prof. Barnard in the Lick Observatory. The discovery has not yet been verified elsewhere.

This new inner satellite appears to be very tiny, only perhaps one hundred miles or so in diameter, and it is seen in the telescope as a star of the thirteenth or fourteenth magnitude. It lies very near to Jupiter, at a distance from that planet's centre of less than 117,000 miles, and it is believed to spin round its huge primary in less than twelve earthly hours — a needful speed to counterbalance the force of gravitation drawing it towards Jupiter.

Some years ago a thought was put forward as to Jupiter's moons. It was suggested that, while Jupiter himself must be regarded as obviously unfitted in his present condition to support living creatures, it might be that the little *moons* were inhabited, and that Jupiter served as a kind of secondary sun to his satellites. Closer observation and growing knowledge have not tended to confirm this notion, and it must now be booked as thrown aside.

The nearer moons have indeed a magnificent view of Jupiter as an enormous bright disc in their sky, shining radiantly with reflected sunlight; and the varying tints and stormy changes in the cloud-belts, viewed from near at hand, must present marvellous effects.

Just as Mercury, Venus, Earth, and Mars travel round the sun at different distances, nearly in the same plane, so do Jupiter's moons travel round him at different distances, nearly in the same plane. The four outer moons are always seen in a line, not one high and another low, one near his pole and another near his equator.

The nearest of the four to Jupiter, named Io, is said to be over two thousand miles in diameter, travels round Jupiter in less than two of our days, and is eclipsed by Jupiter's shadow once in every forty-two hours.

The next moon, Europa, is rather smaller, takes over three days to its journey, and suffers eclipse once in every eighty-five hours.

The next moon, Ganymede, is believed to be considerably larger than Mercury, journeys round Jupiter once a week, and is eclipsed once every hundred and seventy-one hours.

The outer moon, Callisto, is also said to be slightly larger than Mercury, performs its journey in something more than sixteen days, and from its greater distance suffers eclipse less often than the other three.

The nearest of the four is more than two hundred thousand miles from Jupiter; the farthest is more than one million miles off.

The fact of these eclipses, and of the shadow cast by Jupiter's body, shows that though he may give out a great deal of heat, he cannot shine with intrinsic light, or it would prevent any shadow from being thrown by the sunlight.

CHAPTER VIII.

SATURN.

"He hath made everything beautiful in His time."—ECCLES. 3:11.

THE system of Jupiter is a simple system compared with that of Saturn, next in order.

Jupiter has only four or five moons, while Saturn has eight, and in addition to these he has three wonderful rings. Neither rings nor moons can be seen without a telescope, on account of Saturn's great distance from us—more than three thousand times the distance of the moon, or upwards of eight hundred and twenty millions of miles.

Saturn does not equal his mighty brother Jupiter in size, though he comes near enough in this respect to be called often his "twin," just as that small pair of worlds, Venus and Earth, are called "twins." While Jupiter is equal in size to over one thousand two hundred earths, Saturn is equal to about seven hundred earths. And while Jupiter is equal in weight to three hundred earths, Saturn is only equal in weight to ninety earths. He appears to be made of very light materials—not more than three-quarters as dense as water. This would show the present state of Saturn to be very different from the present state of the earth. We are under the same uncertainty in speaking of Saturn as in speaking of Jupiter. Like Jupiter, Saturn is covered with dense masses of varying clouds, occasionally opening and allowing the astronomer peeps into lower cloud-levels, but rarely or never permitting the actual body of the planet to be seen.

The same perplexities also come in here, to be answered much in the same manner.

We should certainly expect that in a vast globe like Saturn the strong force of attraction would bind the whole into a dense solid mass.

Instead of which Saturn is much the least solid of all the planets. He seems to be made up of a very light substance, surrounded by vapor.

One explanation can be offered. What if the globe of Saturn be still in a red-hot molten state, keeping such water as would otherwise lie in oceans on his surface floating aloft in masses of steam, the outer parts of which condense into clouds?

No one supposes that Jupiter and Saturn are in the same condition of fierce and tempestuous heat as the sun. They may have been so once, but they must now have cooled down very many stages from that condition. Though no longer, however, a mass of far-reaching flames and fiery cyclones, the body of each may have only so far cooled as to have reached a stage of dull molten red heat, keeping all water in the form of vapor, and sending up strong rushes of burning air to cause the hurricanes which bear to and fro the vast cloud-masses overhead.

The diameter of the largest moon is about half the diameter of the earth, or much larger than Mercury. The four inner satellites are all nearer to Saturn than our moon to us, though the most distant of the eight is ten times as far away. The inner moon takes less than twenty-three hours to travel round Saturn, and the outer one over seventy-nine days.

A great many charming descriptions used to be worked up, with Saturn as with Jupiter, respecting the magnificent appearance of the eight radiant moons, joined to the glorious shining of the rings, as quite making up for the diminished light and heat of the sun. But undoubtedly

Saturn's cloudy covering would much interfere with observations of the moons by any inhabitants of the solid body within—supposing there were any solid body at all. And though it sounds very wonderful to have eight moons instead of one moon, yet all the eight together give Saturn only about one-sixteenth part of the light which we receive from our one full moon—so much more dimly does the sun light them up at that enormous distance.

One more possible proof of Saturn's half-liquid state is to be found in his occasional very odd changes of shape. Astronomers have been startled by a peculiar bulging out on one side, taking off from his roundness and giving a square-shouldered aspect.

We may not, perhaps, count it impossible that a solid globe should undergo such tremendous upheavals and outbursts as to raise a great portion of its surface five or six hundred miles above the usual level—the change being visible at a distance of eight hundred millions of miles. It is, however, far easier to understand the possibility of such an event in the case of a liquid seething mass than in the case of a solid ball.

On the other hand this alteration of outline

may be caused simply by a great upheaval, not of the planet's surface, but of the overhanging layer of clouds.

Some such changes, only much slighter, have been remarked in Jupiter.

And now as to the rings. Nothing like them is to be seen elsewhere in the Solar System.

They are believed to be three in number, though some would divide them into more than three. Passing completely round the whole body of Saturn, they rise, one beyond another, to a height of many thousands of miles.

The inner edge of the inner ring—an edge perhaps one hundred miles in thickness—is more than ten thousand miles from the surface of Saturn, or more strictly speaking, from the outer surface of Saturn's cloudy envelope. A man standing exactly on the equator and looking up, even if no clouds came between, would scarcely be able to see such a slender dark line at such a height.

This dark transparent ring, described as dusky or purple, and known as the Crape Ring, rises upwards to a height or breadth of nine thousand miles. Closely following it is a ring more brilliant than Saturn himself, over eighteen thousand miles in breadth. When astronomers talk of the "breadth" of these rings it must be understood that they mean the width of the band measured *upwards*, in a direction away from the planet.

Beyond the broad bright ring is a gap of about one thousand seven hundred miles. Then follows the third ring, ten thousand miles in breadth, its outermost edge being at a height of more than forty-eight thousand miles from Saturn. The color of the third ring is grayish, much like the gray marking often seen on Saturn.

At one time it was supposed that these rings were solid, but they are now believed to consist of countless myriads of meteorites, each whirling in its own appointed pathway round the monster planet.

As already said—leaving out of the question the cloudy atmosphere—a man standing on the equator would see nothing of the ring. A man standing at the north pole or the south pole of Saturn could see nothing either, since the rings would all lie below his horizon. But if he travelled southward from the north pole, or northward from the south pole, towards the equator, he would in time see the ringed arch appearing

DONATI'S COMET, 1858.

above the horizon, rising higher and growing wider with every mile of his journey. And when he was in a position to view the whole broad expanse, the transparent half-dark belt below, the wide radiant band rising upwards over that, and the grayish border surmounting all, he would truly have a magnificent spectacle before him.

This magnificent spectacle is however by no means always visible, even from those parts of Saturn where alone it ever can be seen. The rings shine merely by reflected sunlight. Necessarily, therefore, while the sunbeams make one side bright the other side is dark; and not only this, but the rings throw broad and heavy shadows upon Saturn in the direction away from the sunlight.

In the daytime they probably give out a faint shining something like our own moon when seen in sunlight. During the summer nights they shine, no doubt, very beautifully. During the winter nights it so happens that their bright side is turned away; and not only that, but during the winter days the rings, while giving no light themselves to the wintry hemisphere of Saturn, completely hide the sun.

When it is remembered that Saturn's winter—

that is, the winter of each hemisphere in turn—lasts during fifteen of our years, and when we hear of total eclipses of the sun lasting unbroken through eight years of such a winter, with not even bright rings to make up for his absence, we cannot think of Saturn as a tempting residence. The sun gives Saturn at his best only about one-ninetieth of the heat and light that he gives to our earth; but to be deprived of even that little for eight years at a time does indeed sound somewhat melancholy.

Any one looking from one of the nearer moons might have splendid views of Saturn and his rings in all their varying phases.

For Saturn is a beautiful globe wrapped in his changeful envelope of clouds, which, seen through a telescope, are lit up often with rainbow tints of blue and gold, a creamy white belt lying usually on the equator, while around extend the purple and shining and gray rings, sometimes rivalling in bright colors Saturn himself.

The moons of Saturn do not, like those of Jupiter, travel in one plane

CHAPTER IX.

URANUS AND NEPTUNE.

"He hath showed His people the power of His works."—
Psa. 111:6.

TILL the year 1781 Saturn was believed to be the outermost planet of the Solar System, and nobody suspected the fact of two great lonely brother-planets travelling round the same sun at vast distances beyond.

Uranus, nine hundred millions of miles from Saturn. Neptune, one thousand millions of miles from Uranus. No wonder they remained long undiscovered!

Uranus can sometimes be seen by the unaided eye as a dim star of the sixth magnitude. And when he was known for a planet it was found that he *had* been often so seen and noted. Again and again he had been taken for a fixed star, and as he moved and disappeared from that particular spot it was supposed that the star had vanished.

One night when Herschel was busily exploring with a powerful telescope he noticed something which he took for a comet without a tail.

He saw it was no mere point of light like the stars, but had a tiny round disc or face, which could be magnified. So he watched it carefully and found in the course of a few nights that it moved—very slowly certainly, but still it did move. Further watching and calculation made it clear that, though the newly-found heavenly body was at a very great distance from the sun, yet it was moving slowly in an orbit *round* the sun. Then it was known to be a planet and another member of the Solar System.

Everybody supposed that now at least the outermost member of all was discovered.

But a very strange and remarkable thing happened.

Astronomers know with great exactness the paths of the planets in the heavens. They can tell, years beforehand, precisely what spot in space will be filled at any particular time by any particular planet. I am speaking now of their movements round the sun and in the Solar System—not of the movements of the whole family with the sun, about which little is yet known.

Each planet has its own particular pathway; its own particular distance from the sun, varying at each part of its pathway; its own particular speed in travelling round the sun, changing constantly from faster to slower or slower to faster according to its distance from the sun and according to the pull backwards or forwards of other neighboring planets in front or in rear.

For as the orbits of all the planets are ovals, with the sun not in the middle, but somewhat to one side of the middle, it follows that all the planets in the course of their years are sometimes nearer to and sometimes farther from the sun.

The astronomers of the present day understand this well, and can describe with exactness the pathway of each planet. This knowledge does not come merely from watching one year how the planets travel and remembering for another year, but is much more a matter of close and difficult calculation.

It seems to be the will of God to govern commonly things in the heavens as He governs commonly things upon earth, by certain regular laws of working. "Laws of nature" we call them, though "laws of God" would perhaps be a truer expression and reach higher.

There was a great astronomer, named Kepler, who discovered some of the wonderful laws by which God governs the movements of the planets in their orbits. "Kepler's Three Laws" are often spoken about, by which is meant "three laws of God's working in nature discovered by Kepler."

Before these laws and others were understood the movements of the planets were a constant puzzle. Now that they have been clearly grasped, astronomers can not only perceive, as it were, the *plan* of all the planets' journeys, but can draw out beforehand a scheme or sketch, according to these laws, telling exactly what the pathway of any one planet is sure to be.

Many things have to be considered, such as the planet's distance from the sun, the sun's power of attraction, the planet's speed, the nearness and weight of other neighboring planets.

All these questions were gone into and astronomers sketched out the pathway in the heavens which they expected Uranus to follow. He would move in such and such an orbit, at such and such distances from the sun, and at such and such rates of speed.

But Uranus would not keep to these rules. He quite discomfitted the astronomers. Sometimes he went fast when according to their notions he ought to have gone more slowly; and sometimes he went slowly when they would have looked for him to go more fast, and the line of his orbit was quite outside the line of the orbit which they had laid down. He was altogether a perplexing acquaintance and difficult to understand. However, astronomers felt sure of their rules and modes of calculation, so often before tested and not found to fail. They made a guess at an explanation. What if there were yet another planet beyond Uranus, disturbing his motions, now drawing him on, now dragging him back, now so far balancing the sun's attraction by pulling in the opposite direction as to increase the distance of Uranus from the sun?

It might be so. Yet who could prove it? Hundreds of years might pass before any astronomer in his star-gazing should happen to light upon such a dim and distant world. Nay, the supposed planet might be, like Uranus, actually seen, and only be mistaken for a "variable star," shining but to disappear.

There the matter seemed likely to rest. There the matter probably would have rested for a good while had not two men set themselves to conquer the difficulty. One was a young Englishman,

John Couch Adams; the other a young Frenchman, Leverrier—both being astronomers.

Each worked independently of the other, neither knowing of the other's toil. The task which they had undertaken was no light one—that of reaching out into the unknown depths of space to find an unknown planet.

Each of these silent searchers into the sky-depths calculated what the orbit and speed of Uranus would be without the presence of another disturbing planet beyond. Each examined what the amount of disturbance was, and considered the degree of attraction needful to produce that disturbance, together with the direction from which it had come. Each, in short, gradually worked his way through calculations far too deep and difficult for ordinary minds to grasp till he had found just that spot in the heavens where a planet *ought* to be to cause, according to known laws, just such an effect upon Uranus as had been observed.

Adams finished his calculation first, and sent the result to two different observatories. Unfortunately his report was not eagerly taken up. It was, in fact, hardly believed. Leverrier finished his calculation also and sent the result to the

Berlin observatory. The planet was actually seen in England first, but the discovery was actually made known from Berlin first. The young Englishman had been beforehand, but the young Frenchman gained foremost honor.

This, however, was of slight comparative importance. The truly wonderful part of the matter was that these two men could have so reasoned that, from the movements of one lately-discovered planet, they could point out the exact spot where a yet more distant planet ought to be, and that close to this very spot the planet was found.

For when both in England and in Germany powerful telescopes were pointed in the direction named, *there the planet was.*

No doubt about the matter. Not a star, but a real new planet in the far distance, wandering slowly round the sun.

This was indeed a triumph of human intellect. Yet it is no matter for human pride, but rather of thankfulness to Him who gave to man this marvellous reasoning power. And the very delight we have in such a success may humble us in the recollection of the vast amount lying beyond of the utterly unknown.

Perhaps the chief feeling of satisfaction in this

particular discovery may be said to arise from the fact that it gives marked and strong proof of the truth of our present astronomical system and beliefs. Many mistakes may be made and much has often to be unlearned. Nevertheless, if the general principles of modern astronomy were wrong; if the commonly-received facts were a delusion, such complete success could not have attended so delicate and difficult a calculation.

We do not know much about these two outer planets, owing to their enormous distance from us.

Uranus is in size equal to sixty-four earths, and Neptune is in size somewhat larger than Uranus. Both these planets are formed of decidedly heavier materials than Saturn, being about as dense as water.

The size of the sun as seen from Uranus is about one three-hundred-and-ninetieth part of the size of the sun we see. To Neptune he shows a disc only one nine-hundredth part of the size of that visible to us—no disc at all, in fact, but only star-like brilliancy to any such eyes as ours.

The Uranian year lasts about eighty-four of our years; and this, with a very sloping axis, must cause most long and dreary winters, the tiny sun being hidden from parts of the planet during half an earthly lifetime.

Uranus has at least four moons, travelling in very different planes from the plane of the ecliptic. Once it was thought that he had eight, but astronomers have since searched in vain for the other four believed for a while to exist. Neptune has one moon and may possess others not yet discovered.

CHAPTER X.

COMETS AND METEORS.

" It is the glory of God to conceal a thing." Prov. 25:2.

AN interesting discovery has been made. It appears that some sort of mysterious tie exists between comets and meteors.

For a long while this was never suspected. How should it be? The comets so vast in size, the meteors so small and evanescent—how could it possibly be supposed that the one had anything to do with the other?

But supposings often have to give in to facts. Astronomers have gradually become convinced that there certainly is a connection between the two.

Comets and meteor-streams are found to occupy, sometimes at least, the very same pathways in the heavens, the very same orbits round the sun. A certain number of meteor-systems are now pretty well known to astronomers as regularly met by our earth at certain points in her yearly journey. Some of these systems or rings have each a comet belonging to it—not merely journeying near, but actually in its midst, on the same orbit.

Perhaps it would be more correct to say that the meteors belong to the comet than that the comet belongs to the meteors. We tread here on uncertain ground; for whether the meteors spring from the comet or whether the comet springs from the meteors or whether both spring from the same source, cannot yet be quite definitely asserted.

Though we cannot fully explain the kind of connection, yet a connection there plainly is. So many instances are now known of a comet and a meteor-ring travelling together that it is doubtful whether any such ring could be found without a comet in its midst. By-and-by the doubt may spring up whether there ever exists a comet without a train of meteors following him.

Among the many different Meteor-Rings which are known, two of the most important are the so-called August and November systems. Of these two the November system must claim our chief attention.

Not that we are at all sure of these being the

most important meteor-rings in the Solar System. On the contrary, as regards the November ring, we have some reason to think that matters may lie just the other way.

The comet belonging to the November system is a small one, quite an insignificant little comet, only visible through a telescope. We do not of course know positively that larger comets and greater meteor systems generally go together; but to say the least it seems likely. And if the greatness of a ring can at all be judged of by the size of its comet, then the November system must be a third-rate specimen of its kind. It is of particular importance to us merely because it happens to be the one into which our earth plunges most deeply, and which we therefore see and know the best. The August ring is on the contrary connected with a magnificent comet, and may be a far grander system. But our pathway does not lead us into the midst of the August meteors as into those of November. We pass seemingly through its outskirts.

The meteors of the November system are very small. They are believed to weigh commonly only a few grains each. If they were larger and heavier, some of them might find their way to earth's surface only half diminished in their rush through the air; but this they seem never to do.

Doubtless Meteor-Systems exist in the Solar System which our earth does not encounter, containing much larger and heavier meteors. It is well for us that we do not plunge into any such ring, or we might find our atmosphere an insufficient protection.

The last grand display of the November system of meteorites took place in the years 1866 to 1869, being continued more or less during three or four Novembers following. The next grand display is not expected until the year 1899.

For this system—Leonides, as it is called*— seems to have a "time" or "year" of thirty-three and a quarter earthly years.

The shape of its orbit is a very long ellipse, near one end of which is the sun, while the other end is believed to reach farther away than the orbit of Uranus.

A great deal of curiosity has been felt about the actual length and breadth and depth of the

* Because the falling-stars in this display seem all to shoot towards us from a spot in the constellation Leo.

stream of meteorites through which our solid earth has so often ploughed her way.

During many hours at a time lookers-on have watched the magnificent display of heavenly fireworks—not a mere shooting-star here and there as on common nights, but radiant meteors flashing and dying by thousands through the sky. In 1866 no less than eight thousand meteors in two hours and a quarter were counted from the Greenwich Observatory. A natural wonder sprang up in many minds as to the extent of the ring from which they fell.

For not in one night only, but in several nights during three or four years, and that not once only, but once in every thirty-three years, thousands and tens of thousands appear to have been stolen by our earth from the meteor-ring, never again to be restored. Yet each time we touch the ring we find the abundance of little meteors in no wise seemingly lessened.

When speaking of a "ring" of meteors it must not be supposed that necessarily the meteors form a whole unbroken ring all round the long oval orbit. There *may* be no breaks. There *may* be a more or less thin scattering throughout the entire length of the pathway. But the meteors certainly seem to cluster far more densely in some parts of the orbit than in other parts, and it was about the size of the densest cluster that so m'ch curiosity was felt.

Little can be positively known, though it is very certain that the cluster must be enormous in extent. Three or four years running, as our earth, after journeying the whole way round the sun, came again to that point in her orbit where she passes through the orbit of the Leonides, she found the thick stream of meteors still pouring on, though each year lessening in amount. Taking into account this fact, and also the numbers that were seen to fall night after night, and also the speed of our earth, a "rough estimate" was formed.

The length of this dense cluster is supposed to reach to many hundreds of millions of miles. The thickness or depth of the stream is calculated to be in parts over hundreds of thousands of miles, and the breadth perhaps not less than one hundred thousand miles. Each meteor is probably at a considerable distance from his neighbors; but the whole mass of them, when in the near neighborhood of the sun, must form a magnificent sight. And if this be only a third-rate system what must a first-rate system be like? And how many such systems are there throughout the sun's wide domains? The most powerful telescope gives us no hint of the existence of these rings till we find ourselves in their midst.

It may be that they are numbered by thousands, even by millions. The whole of the Solar System—nay, the very depths of space beyond—may, for aught we know, be crowded with Meteor Systems. Every comet may have his stream of meteors following him, but though the comet is visible to us, the meteors are not. Billions upon billions of them may be ever rushing round our sun, entirely beyond our ken, till one or another straggler touches our atmosphere to flash and die as a "shooting star" in our sight.

We have, and with our present powers we can have, no certainty as to all this. But I may quote here the illustration of a well-known astronomical writer on the subject.

Suppose a blind man were walking out of doors along a high-road, and during the course of a few miles were to feel rain falling constantly upon him. Would it be reasonable on his part if he concluded that a small shower of rain had accompanied him along the road as he moved, but that fine weather certainly existed on either side of the road? On the contrary, he might be sure that the drops which he felt were but a few among millions falling all together.

Or look at the rain-drops on your window some dull and rainy day. Count how many there are. Could you, with any show of common-sense, decide that those rain-drops, and those alone, had fallen that day in your town?

So when we find these showers of meteors falling to earth we may safely conclude that, for every one which touches our atmosphere, myriads rush elsewhere in space, never coming near us.

CHAPTER XI.

MORE ABOUT COMETS AND METEORS.

" He commanded, and they were created."—PSA. 148:5.

THERE have been several comets of importance seen in the present century.

In 1811 a remarkably fine one appeared. The bright nucleus was only about four hundred miles in diameter, but the whole head, including the envelope or coma, measured one hundred and twelve thousands of miles across, and the enormous tail stretched out to a distance of one hundred and twelve millions of miles. This comet travels in so long and narrow an ellipse that some say his return must not be looked for in less than thirty centuries.

The great comet which in 1682 received first the name of Halley's Comet appeared last in 1835, his return having been foretold within three days of its actually taking place. For Halley's Comet is a member of the Solar System having a yearly

More About Comets and Meteors. 245

journey of seventy-six earthly years. He journeys nearer to the sun than Venus and travels farther away than Neptune.

In 1843 appeared one of the brightest comets ever seen, and also one of the nearest to the sun. So close ran his orbit to the centre of our system that at one time his nucleus was only about thirty thousand miles from the sun's surface. When we remember that sun-flames often rise to a height of more than fifty thousand miles we shall see how terrific must have been the heat endured.

In the years 1858, 1861, and 1862 three more comets appeared, all visible without the help of a telescope. Of these three, Donati's Comet, in 1858, was far superior to the rest.

It was a singular fact about the comet of 1858 that at one time the star Arcturus could be seen shining through the densest portion of the tail, close to the nucleus. Now, although the faintest cloud-wreath of earth would dim if not hide this star, yet the tail of the comet was of so transparent a nature that Arcturus shone undimmed, as if no veil had come between. The exceedingly slight and airy texture of a comet's tail could hardly be more plainly shown.

It was this gauzy appendage to a little nucleus which men once thought could destroy our solid earth at a single blow!

Yet, while taking care not to overrate, we must not underrate. True, the comets are delicate and light in structure. One comet, in 1770, wandered into the very midst of Jupiter's moons, and so small was its weight that it had no power whatever, so far as has been detected, to disturb the said moons in their orbits. Jupiter and his moons did very seriously disturb the comet, however; and when he came out from their midst, though none the worse for his adventure, he was forced to travel in an entirely new orbit, and never managed to get back to his old pathway again.

But there are comets *and* comets, some being heavier than others. The comet named after Donati, albeit too transparent to hide a star, was yet so immense in size that his weight was calculated by one astronomer to amount to as much as a mass of water forty thousand miles square and one hundred and nine yards deep.

When first noticed, Donati's Comet had, like all large comets, a bright envelope of light round the nucleus. After a while the one envelope grew into three envelopes and a new tail formed beside the principal tail, which for a time was seen to bend gracefully into a curve like a splendid plume. A third but much fainter tail also made its appearance, and many angry-looking jets were poured out from the nucleus. These changes took place while the comet was passing through the great heat of near neighborhood to the sun. Afterwards, as he passed away, he seemed gradually to cool down and grow quiet.

The singular changes in the appearance of Newton's Comet have been earlier noticed. No marvel that he did undergo some alterations. The tremendous glare and burning heat which that comet had to endure in his rush past the sun were more than twenty-five thousand times as much as the glare and heat of the fiercest tropical noonday ever known upon earth. Can we wonder that he should have shown "signs of great excitement," that his head should have grown larger and his tail longer?

It certainly was amazing and bewildering that the said tail, over ninety millions of miles in length, should in four days have seemingly swept round in an enormous half-circle, so as first to point in one direction and then to point in just the opposite direction.

We are much in the habit of speaking about comets as travelling through the heavens with their tails streaming behind them. But though this is sometimes the case it is not always so.

The tails of comets always stream *away from the sun*, whether before or behind the comet's head seeming to be a matter of indifference.

As the comet comes hurrying along his orbit with ever-increasing speed towards the sun, the head journeys first and the tail follows after.

But as the comet rounds the loop of his orbit near the sun—the point nearest of all being called his *perihelion*—the head always remains towards the sun, while the tail swings, or seems to swing, in a magnificent sweep round, pointing always in the direction just away from the sun.

Then, as the comet journeys with slackening speed on the other side of his orbit, towards the distant *aphelion* or farthest point from the sun, he still keeps his head towards the sun. So at this part of his passage, in place of the head going first and the tail following after, the tail goes first and the head follows after. The comet thus appears to be moving backwards. Or, like an engine pushing instead of drawing a train, the head seems to be driving the tail before it.

This careful avoidance of the sun by comets' tails is remarkable. Astronomers speak of the "repulsive energy" with which the sun "sweeps away" from his neighborhood the light gauzy appendages of a comet; but the precise nature of that repulsion and the mode of its working are not yet manifested.

It need not be supposed that the enormous tails, millions of miles in length, are formed of unbroken and connected substance, whirled round, like a vast kite-tail or a stupendous fan, at a perfectly unimaginable speed. The actual make of the tails cannot yet be stated with certainty; but there is a growing belief among astronomers that electricity may have something of a hand in the matter.

After all, the tail is a mere adjunct to the comet—no more needful to a comet's existence than a man's beard is to the man's existence. According to ordinary notions, a full-blown comet consists of the bright central nucleus, the surrounding coma, and the far-reaching tail. But if you have the nucleus you have the comet, and no more is necessary. It is known to be a comet, not a star, mainly by the path that it follows and the rates of its motions.

The coma has been described by Dr. Huggins as "usually a luminous fog, surrounding the nucleus and gradually shading off from it;" and the tail as "a continuation, in a direction opposite to that of the sun, of the luminous fog of the coma."

Except in the sun's neighborhood a comet's tail has commonly no existence. As the wanderer draws near to our luminary, a tail of less or greater dimensions is often developed under the increasing heat: and as the wanderer passes away once more the tail diminishes and disappears.

Spectrum analysis declares that comets usually consist of carbon, hydrogen, nitrogen, and possibly oxygen—all, of course, in a state of highly-heated and light-emitting gas. To some extent also a comet reflects the sunlight.

More than this is known of the nature of comets through studying the nature of their near of kin—meteors and meteorites. Speaking at the same time as just cited, Dr. Huggins said further, "We are thus led to see the close physical connection and oneness of origin, if not indeed identity of nature, of comets and of these meteors. Now, the meteors on these occasions"—the great meteor showers — "are too minute to pass through the ordeal of ignition by our atmosphere: they are burned up before they reach the earth: but at other times small celestial masses come down to us which, there can be little doubt, are of the same order of bodies and similar in chemical nature."

Here again spectrum analysis steps in and tells us that in the make of celestial meteorites falling to earth, among divers other substances may be found the hydrogen, carbon, and nitrogen which we have already seen to be the main substances in the composition of comets.

True, a fallen meteorite is small, solid, heavy, while a comet is light and hazy. But if the materials of which a meteorite is made were rendered gaseous, then at once it would be not only much larger, but light and hazy.

In 1867 a large lecture-room was entirely lighted by means of gas extracted from a meteorite which had arrived from distant space.

Not only do solid aerolites fall half-burned to the ground, but even when the meteorites are quite consumed in the air the fine dust remaining still sinks earthward. This fine dust has been found upon mountain-tops, and has been proved by close examination to be precisely the same as the material of the solid aerolites.

It is a wonderful thought that we should really have these visitants from the sky, solid metal or showers of dust coming to us from distant space.

If the dust of thousands of meteorites is always thus falling earthward, one would imagine that it must in time add something to the weight of the earth. And this actually is the case. During the last three thousand years no less than one million tons of meteorite-dust must, according to calculation, have fallen to earth out of the sky. A million tons is of course a mere nothing compared with the size of the world. Still the fact is curious and interesting.

Some of the tremendous outbursts seen on the surface of the sun have been described earlier. It is believed that in such outbursts matter is driven forth with violence sufficient to send it whirling through space, never to return thither. Many meteorites may have had their birth thus, whether from our own sun or from stars like in nature to our sun.

It has been conjectured that some meteorites *may* also have sprung from earthly volcanoes, being shot aloft with impetus enough to carry them to a vast distance, breaking loose perhaps for a while from earth's control, but after a lengthened period—it may be accidentally — travelling earthward once more.

A few more words about "comet-visitors."

Many comets, as stated earlier, belong to the Solar System; but many also come only once, flashing round the sun and rushing away in a new direction, never to return. Any comet at about the distance of our earth from the sun, travelling at a rate of more than twenty-six miles per second, is beyond solar control. The sun is powerless to hold prisoner so impulsive a vagrant, and it *must* find its way out of the Solar System.

Comets which come to us from outside our System must come from some other System. And the nearest Systems known are those of the stars.

The nearest star of all, among those whose distance has yet been measured, is Alpha Centauri. A comet passing from Alpha Centauri to our Solar System would require millions of years for his journey. But it is not impossible that other stars may lie nearer to us than Alpha Centauri, although the fact has not yet been discovered.

END OF PART II.

PART III.

CHAPTER I.

MANY SUNS.

"I, even my hands, have stretched out the heavens, and all their host have I commanded." Isa. 45 : 12.

ONCE more we have to wing our flight far far away from the busy Solar System where we live; away from whirling planets, moons, meteorites, all shining with reflected sunlight; away from the great central sun himself, our own particular bright star. Once more we have, in imagination, to cross the vast black empty space—*is* it black? and *is* it empty? had we sight to see things as they are—separating our sun from other suns, our star from other stars.

For the sun is a star—only a star. And stars are suns—big blazing suns. One is near and the others far away, that is the main difference.

We have no longer to do with bodies merely reflecting another's light—always dark on one side and bright on the other—but with burning bodies, shining all around by their own light. We have no longer to picture just one single star with his surrounding worlds, but we have to fix our thoughts upon the great universe of stars or suns in countless millions.

The sun is centre and ruler and king in his own system. But as a star he is only one among many stars, some greater, some less than himself.

Thought travels fast, faster than a comet, faster than light. A rushing comet would, it is believed, take eight millions of years to cross the chasm between the nearest known fixed star and us. Light, flashing along at the rate of about 186,300 miles a second, will perform the same journey in four years and one-third. But thought overleaps the boundary in less than four minutes.

Each star that we see in the heavens is to our eyesight simply one point of light. The brighter stars are said to be of greater magnitude, and the fainter stars of lesser magnitude, yet one and all they have no apparent size. The most powerful telescope, though it can increase their brilliancy, cannot add to their size. A planet which to the naked eye may look like a star will, under a telescope, show a disc the breadth of which can be measured or divided; but no star has any real sensible disc in the most powerful telescope yet constructed.

The reason of this is the enormous distance of the stars. Far off as many of the planets lie, yet the farthest of them is as a member of our household compared with the nearest star.

I have already tried to make clear the fact of their vast distance. Light, which comes to us from the sun in less than nine minutes, takes four years and four months to reach us from Alpha Centauri. From this four-years-and-a-half length of journey between Alpha Centauri and Earth the numbers rise rapidly to twenty years, fifty years, hundreds of years, even thousands of years. The distances of most of the stars are completely beyond our power to measure. The whole orbit of our earth, nay, the whole wide orbit of the far-off Neptune, would dwindle down to one single point, if seen from the greater number of the stars.

It used to be believed that, taking the stars generally, there was probably no very marked difference in their size, their kind, their brightness. Some of course would be rather larger and others rather smaller; still it was supposed that they might be roughly classed as formed much on the same scale and the same plan. But of late years this notion has been utterly given up.

A very similar idea used to be held with regard to the Solar System. The wonderful variety of form and richness in numbers now known to abound within its limits are discoveries of later years; and now the same variety in kind and size is found also among the stars. The more we look into the heavens the more we see that dull blank uniformity is not to be found there.

It is the same upon earth. Man builds his little rows of box-like houses side by side, each one exactly like all the rest, or dresses his thousand soldiers in coats of the same cut and color, or repeats a neat leaf-design hundreds of times on carpets or wall-papers. But God never makes two leaves or two blades of grass alike. Wholesale turning out of things after one pattern is quite a human idea, not divine.

We know so much about the stars as that some are at least considerably larger and some considerably smaller than others.

When one star is seen to shine brightly and another beside it shines dimly, we are apt to think that the brightest must be the nearest. Yet it is often impossible for us to say how much of the difference is owing to the greater distance of one or the other, to the greater size of one or the other, or to the greater brilliancy of one or the other.

In many instances we do know enough to be quite sure that there is a great difference, not only in the distance of the stars, but in their size, their kind, their brightness.

The stars have been sometimes classed into four distinct orders or degrees—partly depending on their color.

The first class is that of the White Suns. These are generally held to be the grandest and mightiest of all. The star Sirius belongs to the order of White Suns.

Secondly comes the class of Golden Suns. To these blazing orbs of yellow light, said to be second only to the white-light stars, belongs our own sun.

Thirdly, there are numbers of stars called Variable Stars, the light of which is constantly changing, now becoming greater, now becoming less.

Fourthly, there is the class of small Red Suns, about which not so much is known.

These four orders or divisions do not by any means include all the stars, or even all the single stars. Roughly speaking, however, the greater number of single stars, and many also of the double stars, belong to one or another of the above classes.

When we talk of the different sizes of the different stars it should be plainly understood that we have no means of directly measuring them. A point of light showing no disc, no surface, no breadth, cannot be measured, for there is nothing to measure.

In certain cases we are not entirely without the power of judging. The distances of some of the stars from us have been found out. Knowing how far off any particular star is, astronomers are able to calculate exactly how bright our own sun would look at that same distance. If they find that our sun would shine just as the star in question shines, there is some reason for supposing that our sun and yonder star may be of the same size. If our sun would shine more brightly than the star shines, there is some reason for supposing that the star may be smaller than our sun. If our sun would shine more dimly than the star shines, there is some reason for supposing that the star may be larger than our sun.

Other matters, however, have to be considered. Suppose we find a star at a certain distance shining twice as brilliantly as our own sun would shine at that same distance. Naturally then we say that star must be much larger than our sun.

The reasoning may be mistaken. We do not know the fact. What if, instead of being a much *larger* sun, it is only a much *brighter* sun?

This is more than possible. Some doubts have been, indeed, expressed whether the surface of *any* star *could* exceed that of our sun in radiance. But we have really no reason to suppose that our sun is a king of suns in brilliance, since certainly he is not so in size or in speed.

When we picture to ourselves the star-depths, the boundless reaches of heavenly space with these countless blazing suns scattered broadcast throughout, we have not to picture a universe in repose.

On the contrary, all is life, stir, energy. Just as in the busy whirl of our Solar System no such thing as rest is to be found, so also it seems to be in the wide universe.

Every star is in motion. "Fixed" as we call them, they are not fixed. Invisible as their movements are to our eyes, through immensity of distance, yet all are moving. Those silent, placid, twinkling specks of light are, in reality, huge roaring, seething, tumultuous furnaces of fire and flame, heat and radiance.

Each, too, is hurrying along his appointed pathway in space. Some move faster, some move more slowly. One mile per second, ten miles per second, twenty, fifty, hundreds of miles per second—thus varying are their rates of speed.

But whether fast or whether slowly, still onward and ever onward they press. Some are rushing towards us, some are rushing away from us. Some are speeding to the right, some are speeding to the left.

Where are they going? Does any single star ever return to his starting-point—wherever that starting-point may have been? Do they journey in vast circles or ellipses round some far-distant centre? What controls them all? Is it the mighty power of some such centre, or does each star by his faint and distant attraction help to control all his brother-stars, to guide them on their appointed path, to preserve the delicate balance of a universe?

How little we know about the matter!

Only so much we can tell—that the controlling and restraining hand of God is over the whole.

Whether by the attraction of one great centre or by the united influences of a thousand fainter attractions, He steers each radiant sun upon its heavenly path, "upholding all things by the word of His power." There is no blundering, no confusion, no entanglement. All is perfect order, calm arrangement, restrained energy.

CHAPTER II.

SOME PARTICULAR SUNS.

"The glory of the LORD shall endure for ever: the LORD shall rejoice in his works."—PSA. 104:31.

IN the constellation of the Swan there is a little dim sixth-magnitude star scarcely to be seen without a telescope. This star, 61 Cygni by name, is the first whose distance from us it was found possible to determine.

We may think it strange that so faint a star was even attempted. Would not astronomers have naturally supposed it to be one of the farther-distant stars?

No, they did not. For though 61 Cygni showed but a dim light, yet his motion—not the daily apparent motion, but the real motion as seen from earth—was found to be so much more rapid than the motion of most other stars that they rightly guessed 61 Cygni to be a rather near neighbor of ours.

Do not misunderstand me when I speak of "more rapid motion" and of "rather near neighborhood."

Some Particular Suns. 267

The real rate of 61 Cygni's rush through space is believed to be about thirty-six miles each second, or more than one thousand millions of miles each year. All we can perceive of this quick motion is that in the course of three hundred and fifty years 61 Cygni travels over a space in the sky about as long as the breadth of the full moon. Little enough, yet far beyond that detected in the greater number of even the brightest stars.

Then again as to the near neighborhood of this star, 61 Cygni is near enough to have his distance measured, and that is saying a good deal. Alpha Centauri, the nearest star of which we know in the southern heavens, is more than two hundred thousand times as far distant as the sun. But 61 Cygni, the nearest star of which we know in the northern heavens, is nearly twice as far, or some forty billions of miles away.

We call 61 Cygni a star, for so he appears to common observers. In reality, instead of being only one star, the speck of light which we call 61 Cygni consists of *two stars*. The two are separated by a gap about half as wide again as the wide gap between the sun and Neptune. Yet so great is their distance from us that to the naked eye the two seem to be one.

These two suns might together make a sun perhaps about one-third as large as our sun. They differ in size, the quicker movements of one showing it to be the smaller; and it is by means of their known distance from one another, and their known rate of motion, that their size, or rather their weight, can be roughly calculated. Of course neither of the two shows any actual measurable disc.

So much and so little is with tolerable certainty known about this particular pair of suns.

Next let us turn to Alpha Centauri—named *Alpha*, the first letter of the Greek alphabet, because it is the brightest star in the constellation of the Centaur. The second brightest star in a constellation is generally called Beta, the third Gamma, the fourth Delta, and so on; just as if we were to name them A, B, C, D, in order of brightness.

The constellation Centaur lies in the southern heavens, close to the beautiful constellation called the Southern Cross, and is invisible in England.

Of all the stars shining in the heavens round our earth, two only—Sirius and Canopus—show greater brilliancy than Alpha Centauri.

As in the case of 61 Cygni, astronomers were led to attempt the measurement of Alpha Centauri's distance by noticing how much more distinct were his movements than the movements of other stars, though less rapid both to the eye and in reality than those of 61 Cygni.

The distance of Alpha Centauri from us is more than two hundred thousand times the distance of the sun, or about twenty millions of millions of miles. In other words, light, travelling at the rate of 186,300 miles each second, takes four years to journey from Alpha Centauri to Earth.

And this, so far as we yet know, is our sun's nearest neighbor in the heavens outside his own family circle.

Alpha Centauri, like 61 Cygni, is found to consist, not of a single star, but of a pair of stars. It is a two-sun system—whether or no surrounded by planets cannot be told.

The two suns of the Alpha Centauri double-star are separated by a distance about twenty-two times as great as the distance of the earth from the sun, yet to the naked eye they show as a single star. Here again one is much smaller than the other; and the smaller revolves round the larger in about eighty-five years.

It is supposed that the two together might form a sun considerably larger and heavier than our sun. This belief is strengthened by the great brilliancy of Alpha Centauri. Our own sun, placed at that distance from us, would shine only about one-third as brightly as he does.

Turning now from the sun whose distance was first measured and from the nearest star with which we are acquainted, let us think about the most radiant star in the heavens—Sirius, "the blazing Dog-star of the ancients," named by one astronomer "the king of suns."

First, as to the color of Sirius. He belongs to the order of "White Suns," and among all the white suns known to us Sirius ranks as chief. There may be many at greater distances far surpassing him in size and weight and brilliancy; but we can only speak so far as we know.

One ancient writer spoke of Sirius as "red" in hue, and others followed suit; but the original epithet was probably a copyist's mistake. It is at all events not to be depended on.

Secondly, as to the distance of Sirius.

Like a few other stars, Sirius lies not quite so far away as to be beyond reach of measurement.

Some Particular Suns. 271

No base-line upon earth would cause the slightest seeming change of position in him, but as our earth journeys round the sun the line from one side of her orbit to the other is found wide enough. A base-line of one hundred and eighty-five millions of miles does cause just a tiny seeming change.

It is very little even with Alpha Centauri, and with Sirius it is much less. The "displacement" of Sirius is so slight that to measure his distance with exactness is impossible. Roughly calculated, Sirius lies at a distance of about fifty billions of miles. Our sun, at the distance of Sirius, would shine merely as a star of the third magnitude. Light, which reaches us from the sun in nearly nine minutes, and from Alpha Centauri in four years and one-third, cannot reach us from Sirius in much less than twenty years.

Thirdly, as to the size of Sirius.

Here, of course, we are in difficulties. Radiantly as Sirius shines on a clear night, and dazzling as he looks through a powerful telescope, he shows no real disc or round surface capable of being measured. But although we cannot measure his size, we do know something of his weight and of his extraordinary brilliance. The brightness of Sirius is so much greater than the brightness of our sun would be at that same distance that it certainly seems to point to the fact of Sirius being very much the larger sun of the two.

The only reliable method by which a star's probable weight can be calculated is through the discovery of a companion to the star and a knowledge of their relative distances, their times of revolution, and the speed of either. For a long while, since Sirius was not known to possess a companion, this mode of computation could not be followed.

Like other suns Sirius has his "proper motion," as it is called—an outward whirl through space of more than one thousand miles each minute: sometimes more, sometimes less. Mysterious irregularities in the pace of Sirius were noted as long ago as the year 1844, and by dint of careful thought a decision was reached that some large disturbing orb must surely belong to Sirius and revolve round him, a body large enough and near enough to exercise a disturbing influence over the great sun, now hastening him on, now pulling him back, even as Jupiter and Saturn by turns hurry and delay one another.

There the matter rested during fully eighteen years. Then one night the dim gleam of this companion was actually seen; and having been once discovered, it was perceived again and again. So the truth of the surmise reached nearly twenty years earlier through simple exercise of reasoning power was proved beyond question.

Although this faintly-shining second star is so near to Sirius as greatly to affect the movements of the latter, disturbing his regularity of speed to an extent which could be noted forty billions of miles away—yet the distance between Sirius and his companion is somewhere about thirty-seven times as great as our earth's distance from the sun. The dim companion appears to be in weight about equal to our sun, while the bright star, Sirius, is not much more than twice as heavy.

Weight, however, tells little as to size. The blazing radiance of Sirius is extraordinary compared with the dimness of his companion sun. Probably the latter is far cooler and consequently far denser and smaller in make: while the dazzling Sirius has doubtless a photosphere of incandescent gases and glowing molten clouds, like to that of our sun, but greatly exceeding it either in size or in actual brilliance, or probably in both.

Fourthly, as to the motions of Sirius, already touched upon.

It was long possible only to observe the movements of a star when he journeyed *sideways* across the sky. A star coming directly towards us or passing straight away from us would always appear to be at rest. Now, however, by means of the spectroscope it has become possible to perceive and measure the "end on" motions of stars to and from our Solar System as well as their sideway journeyings.

The sideway motion of Sirius had been discovered before, but it now appears that he does not move exactly sideways. He rushes away in a slanting direction at the rate of over one thousand millions of miles each year. Think how many millions upon millions of millions of miles he must now be farther off from us than in the years of the ancients! Yet he shines on with undimmed radiance, still the most brilliant star in our sky — so small a matter are all those millions compared with the sum of his vast distance from earth!

It is an interesting fact that while Sirius is moving away from us, we are also moving away from him. Our "first station ahead," which seems to lie in or near the constellation of Hercules, stands just in the opposite direction from the goal of Sirius.

Fifthly, has Sirius a family or a system of worlds like our sun?

He may have, or he may not. One companion he does possess, in the shape of a companion sun, greatly inferior to himself in lustre. Whether those two suns possess a family of revolving worlds we have no power to say. If one asks, Why not? no reason to the contrary can be given. But to say that a thing may be is not to say that it is. Any star in the heavens *may* have such a family, and any star may pursue his lonely path without an attendant near.

When we do not and cannot know, it is worse than useless to be positive. Still, the discovery after long ages of this dim huge companion to Sirius, and the knowledge that a very slight further increase of dimness, or decrease, rather, of shining, would have made that companion absolutely undiscoverable by us, is full of significance. Even a faint star at that distance can barely be detected; and a planet, shining by borrowed light, could not possibly become visible. Such bodies may exist in countless billions throughout space where we see only the glimmer of a more radiant sun dotting here and there the vast expanse. The mere possibilities thus opened out to imagination are almost appalling!

Lastly, what is Sirius made of?

Here again the spectroscope comes to our aid. Before the discovery of Spectrum Analysis, and its application to astronomy, astronomers can scarcely be said to have known with certainty that stars were suns; and of the composition of those stars, nay, even of our own sun, they were in ignorance.

Now we have learned much! Now the spectroscope, by breaking up and analyzing the slender ray of light which travels earthward from each star, has shown us something of the true nature of those far-off lamps of heaven. Now we know that most of the stars—not all of them—are, like our sun, "great radiating machines" for the giving forth of light and heat and power. Like our sun each is surrounded by a brilliant photosphere, formed in a fiery atmosphere of heated and glowing gases.

In the White Suns, of which Sirius is king and chief, while sodium, magnesium, and other metals do more or less exist, the principal component of this dazzling envelope is found to be incandescent hydrogen.

Our sun belongs, not to the first or Sirian Class of White Suns, but to the second or Solar type of Golden Suns. In them, though hydrogen is abundant, it is not so absolutely prominent.

CHAPTER III.

DIFFERENT KINDS OF SUNS.

"Great things doeth He, which we cannot comprehend."—
JOB 37 : 5.

VARIOUS in kind, various in size, various in color, various in position, various in motion, are the myriad suns scattered through space.

So far are they from being formed on the same plan, turned out on the same model, that it may with reason be doubted whether any two stars could be found exactly alike. Why should we expect to find them so? No two oak-leaves, no two elm-leaves, precisely alike are to be found upon earth.

So some stars are large, some are small. Some are rapid in movement, some are slow. Some are yellow, some white, some red, green, blue, purple, or gray. Some are single stars; others are arranged in pairs, trios, quartettes, or groups. Some appear only for a time, and then disappear altogether. Others are changeful, with a light that regularly waxes and wanes in brightness.

We have now to give a little time and thought to Variable Stars and Temporary Stars; afterwards to Double Stars and Colored Stars.

There are many stars which pass through gradual and steady changes, first brightening, then lessening in light, then brightening again.

One such star is to be seen in the constellation of the Whale. It is named " Mira," or the " Marvellous," and the time in which its changes take place extends to eleven months. For about one fortnight it is a star of the second magnitude. Through three months it grows slowly more and more dim, till it becomes invisible not only to the naked eye, but through ordinary telescopes. About five months it remains thus. Then again during three months it grows brighter and brighter, until it is once more a second-magnitude star, and after a fortnight's pause begins anew to fade.

There is also a variable star in the constellation of Perseus named Algol. This too waxes and wanes regularly, but much more rapidly, since it runs through its list of changes in less than three days. It never becomes brighter than a star of the second magnitude or dimmer than a star of the fourth magnitude.

Another variable star, Betelgeuse, in Orion, undergoes its variations in about two hundred days; while yet another, Delta Cephei, takes only six days.

Our own sun has been reckoned as in some slight degree a variable star, passing through his changes in eleven years. When the sun-spots are most numerous he might appear, possibly, if seen from a great distance, slightly more dim than when there are none of them.

Sometimes a new or Temporary Star appears, blazing brightly for a brief space and then vanishing. It may be a dim star, already known, which thus springs into a new and radiant phase of being; or it may be, practically to us, a new existence. But doubtless the star was there before, a body too faint for us to perceive; and doubtless it is there still when the fires of its outburst die away, once more too faint for its shining to reach our eyes.

An extraordinary specimen of a temporary star was seen in 1572. It was not a comet, for it had no coma or tail, and it never moved from its place. The brightness of the star was so great as to surpass Sirius and Jupiter and to equal Venus at her greatest brilliancy. Nay, it must have surpassed even Venus, for it was plainly visible at mid-day in a clear sky. Gradually the light faded and grew more dim, till it became a mere faint star. As it lessened in brilliancy it also changed in color, passing from white to yellow and from yellow to red. This succession agrees with the three tints of the first, second, and fourth orders of suns as lately classified.

Many other instances have been known, beside those referred to, of variable and temporary stars.

There are two distinct kinds of double-stars. First we have those which merely *seem* to be double, because one lies almost directly behind the other, though widely distant from it—just as a church-tower two miles off may appear to stand close side by side with another church-tower two-and a half miles off, though they are in fact separated. Secondly, we have the real systems of two suns belonging to one another, the smaller moving round the larger, or more correctly both travelling round one central point called the centre of gravity, the smaller having the quicker rate of motion.

Alpha Centauri and 61 Cygni have been already described as examples of true double-stars.

In the constellation Lyræ a marked instance is to be seen. The brightest star in the Lyre is Vega, and near Vega shines a tiny star which to people with particularly clear sight has sometimes rather a longish look.

If you examine this star through an opera-glass, you will find it to consist of two separate stars.

But if you get a more powerful telescope and look again, you will find that *each star* of the couple actually consists of *two* stars. The four are not at equal distances. Two points of light seemingly close together are parted by a wide gap from two other points equally close together. These four stars are believed to have a double motion. Each of the separate pairs revolves by itself, the two suns travelling round one centre; and in addition to this the two *couples* of suns probably perform a long journey round another centre common to them all.

Many thousands of double stars have been discovered; and a large number of these are now known to be not merely two distinct suns lying in the same line of sight, but two brother-suns, each probably the centre of his own system of planets.

We have not only to consider the number of suns, though of simple numbers more yet remains to be said: attention must also be given to the varying colors of different stars.

For all suns in the universe are not made after the model of our sun. All suns are not yellow.

So far as single stars are concerned, colors seem rather limited. White stars, golden or orange stars, ruby-red stars, placed alone, are often seen; but blue stars, green stars, gray stars, silver stars, purple stars, are seldom if ever visible to the naked eye or known to exist as single stars.

Take a powerful telescope and examine star-couples, and a very different result you will find. Not white, yellow, and red alone, but blue, purple, gray, green, fawn, buff, silvery white, and coppery hues will delight you in turn.

As a rule, when the two stars of a couple are alike in color they are either white or yellow or red. Also in the case of double stars of different colors, the larger of the two is almost invariably white or some shade of yellow or red.

There are, however, exceptions to all such rules. Blue stars are almost never seen alone, and as one of a pair the blue star is generally, if not invariably, the smaller. But instances are known of double-stars both of which are blue; and one group in the southern heavens is entirely made up of a multitude of bluish suns.

It is when we come to consider double-stars of two colors that the most striking effects are found.

Now and then the two suns are nearly the same in size, but more commonly one is a good deal larger than the other. This is known by the brighter light of the larger and the more rapid movements of the smaller. The lesser star is often only small by comparison, and may be in reality a very goodly and brilliant sun.

Among nearly six hundred "doubles" examined by one astronomer there were three hundred and seventy-five in which the two stars were of one color, generally white, yellow, orange, or red. The rest were different in tint—the difference between the two suns in about one hundred and twenty cases being very marked.

For instance, a red "primary," as the larger star is called, will be seen with a small green satellite; or a white primary will have a little brother-sun of purple or of dark ruby or of light red. Sometimes the larger sun is orange, the companion being purple or dark blue. Again, the chief star will be red with a blue satellite, or yellow with a green satellite, or orange with an emerald satellite, or golden with a reddish-green satellite. We hear of golden and lilac couples, of cream and violet pairs, of white and green companions. But indeed the variety is almost endless.

There may be worlds circling round these suns — worlds perhaps with living creatures on them. We know little about how such systems of suns and worlds may be arranged. Probably each sun would have his own set of planets, and both suns with their planets would travel round one central point. Perhaps, where the second sun is much the smallest, it might occasionally be like a big blazing satellite among the planets.

Among colored stars, single and double, a few may be mentioned by name as examples.

Sirius, as already observed, is a brilliant white sun; and brilliant white also are Vega, Canopus, Regulus, Spica, and many others.

Capella, Procyon, the Pole-star, and our own sun are examples of yellow stars.

Aldebaran, Betelgeuse, and Pollux are ruby-red.

Antares is a red star with a greenish "scintillation" or change of hue in its twinkling. A tiny green sun belonging to this great and brilliant red sun has been discovered. Antares is sometimes called "the Sirius of Red Suns."

The two double-stars, 61 Cygni and Alpha Centauri, are formed each of two orange suns.

In the Southern Cross there is a wonderful group of stars, consisting of about one hundred and ten suns, nearly all invisible to the naked eye. Among the principal stars of this group, which Sir John Herschel described as being, when viewed through a powerful telescope, like "a casket of variously colored precious stones," are two red stars, two bright green, three pale green, and one of a greenish-blue.

CHAPTER IV.

GROUPS AND CLUSTERS OF SUNS.

"He telleth the number of the stars; He calleth them all by their names."—PSA. 147:4.

WE have been thinking a good deal about single stars and double stars as seen from earth. Now we have to turn our attention to groups, clusters, *masses* of stars in the far regions of space.

Have you ever noticed on a winter night, when the sky was clear and dotted with twinkling stars, a band of faintly-glimmering light stretching across the heavens from one horizon to the other?

The band is irregular in shape, sometimes broader, sometimes narrower, here more bright, there more dim. If you were in the southern hemisphere you would see the same soft belt of light passing all across the southern heavens. This band or belt is called the Milky Way. Sometimes it is called "the Stellar System," or "the Universe." Early in this book mention was made of the word "universe" as used in two senses. The more modern use of it for the Milky Way, or for the particular great galaxy or gathering of stars to which our sun belongs, is the smaller meaning.

But what is the Milky Way?

It is made up of stars. So much we know. As the astronomer turns his telescope to the zone of faintly-gleaming light he finds stars appearing behind stars in countless multitudes; and the stronger his telescope, the more the white light changes into distant stars.

Our sun we believe to be one of the stars of the Milky Way, merely one star among millions of stars, merely one golden grain among the millions of sparkling gold-dust grains scattered lavishly through creation—scattered not recklessly, not by chance, but placed, arranged, and guided each by its Maker's upholding hand.

The Milky Way, the Galaxy, or the Universe, as it has been variously called, has great interest for astronomers. Many have been the attempts made to discover its actual size, its true shape, how many stars it contains, how far it extends. But to all such questions the only safe answers to be returned are fenced around with "perhaps" and "may be."

There are many very remarkable clusters of stars to be seen in the heavens — some few visible as faint spots of light to the naked eye, though the greater number are only to be seen through a telescope. Either with the naked eye or in telescopes of varying power they show first as mere glimmers of light, which, viewed with a more powerful telescope, separate into clusters of distant stars.

The most common shape of these clusters is globular — to the eye appearing simply round. Stars gather densely near the centre and gradually open out to a thin scattering about the edge. Thousands of suns are often thus collected into one cluster.

The clusters are to be seen in all parts of the sky, but the greater number seem to be gathered into the space covered by the Milky Way and by the famous south Magellanic Clouds.

Some of them are beautifully colored, as, for instance, a cluster in Toucan, not visible from England, the centre of which is rose-colored, bordered with white. No doubt it contains a large number of bright red suns surrounded by a scattering of white suns.

By some it is supposed that many of these clusters *may* be other vast galaxies of stars, more or less like the Milky Way, lying at enormous distances from us.

Others, on the contrary, rather maintain that such clusters, instead of being separate "universes" from our own, are merely lesser star-systems, included within our great Stellar System, and probably all travelling round some one centre.

For the present this must be considered an open question. But in either case if worlds are journeying among the suns of such a starry cluster, what sights might not be witnessed by their—possible—inhabitants!

We do not indeed know the distances between the separate suns of a cluster, which may be far greater than appears to us. Yet, on the other hand, it may well be that they are near enough together to shed bright light on all sides of a planet revolving in their midst. The planet might or might not have within view a single sun equal in apparent size to our sun as seen from earth; yet thousands of lesser suns shining night and day would cause a radiance which we never enjoy.

No, not *night*. In such a world there could be no night. Worlds in the midst of a star-cluster must be regions of perpetual day. No night, no starry heaven, no sunrise lights or sunset glories, no shadow mingling with sunshine, but one continual ceaseless blaze of brightness. We can hardly picture even in imagination such a condition of things.

Beside star-clusters there are also Nebulæ.

The word "nebula" comes from the Latin word for "cloud," and the nebulæ are so named from their hazy and cloud-like appearance.

It is not always easy at first sight to distinguish between nebulæ and very distant star-clusters. Both have the same dim and cloudy look. In days gone by the star-clusters were often included by astronomers under the general term of nebulæ. And as with star-clusters, so with many nebulæ, more powerful telescopes have shown them to be great systems or collections of innumerable stars crowded together by vastness of distance.

But although some nebulæ are really, to a certain extent, much the same as star-clusters, others are entirely different. The spectroscope, by its delicate test, proves them to consist of enormous masses of glowing gas, and not to be solid bodies at all, even in the limited sense in which we may call our sun a "solid" body. Whether they are on their road to become solid bodies by-and-by, is another question. Certainly they are not so yet.

By glowing gas I do not mean *burning* gas, like the gas alight at a chandelier burner. For actual burning—for being consumed and wasting away—an atmosphere such as ours is necessary. The gases which compose the outer envelopes of sun and stars, and the gases which compose the whole body of many a nebula, shine with intensity of glowing but non-consuming heat; and this is termed *incandescence*.

The number of nebulæ now known amounts to thousands. They are usually divided into classes, according to their apparent shapes, as seen from earth. There are circular nebulæ, oval nebulæ, annular nebulæ, conical nebulæ, cometary nebulæ, spiral nebulæ, and nebulæ of divers other descriptions. These apparent shapes would doubtless change greatly if we could look upon them from some other standpoint and at a lessened distance. Even when seen in more powerful telescopes, they often wear a totally different aspect.

Many nebulæ are believed to lie at incalculable

distances: and the light which comes to us from them may have been hundreds of years upon its road. But opinions differ much on this question. Some astronomers place many of the nebulæ, as a matter of probability, far outside the vast Stellar System to which we belong; while others believe that most if not all of the gaseous nebulæ are included within the bounds of our own particular starry universe.

Mention must be made of the great nebula in Orion. The middle star of Orion's sword-handle, when carefully noted with the naked eye, can be seen to have a slightly misty aspect. This, seen in powerful telescopes, develops into a glorious nebula, one of the grandest in our heavens. It is shown by the spectroscope to consist partly of stars, partly of glowing gas—of countless suns, possibly not large in size, but great in numbers, "bathed in and surrounded by a stupendous mass of glowing gas!"*—hydrogen gas, nitrogen gas, and one other gas, the nature of which we do not know.

As for the size of this nebula, we are driven to conjecture; but undoubtedly it is enormous. Measurement of size is impossible where measure-

* R. S. Ball.

ment of distance cannot be accomplished. It has, however, been suggested that, if a ball be imagined so huge as to fill the entire circle of our earth's annual pathway round the sun, one million of such vast globes of light might be less in size than the nebula in Orion. How much larger yet it may be none can venture to say.

Another famous nebula is that in Andromeda. It can be seen by the naked eye and has been sometimes taken for a comet. The spectroscope shows this nebula to be formed seemingly, not of glowing gas, but of multitudinous faint stars: so it is more akin to an ordinary star-cluster.

Before quitting this subject allusion should be made to the famous Magellanic Clouds in the southern heavens. Sometimes they are called the Cape Clouds.

They differ from other nebulæ in many points, and more particularly in their apparent size. The Great Cloud is about two hundred times the size of the full moon, while the Small Cloud is about one quarter as large. In appearance they are not unlike two patches of the Milky Way, separated and moved to a distance from the main stream.

These clouds are surrounded by a very barren portion of the heavens, containing few stars; but in themselves they are peculiarly rich. Seen through a powerful telescope they are found to abound with stars. The Greater Cloud alone contains over six hundred from the seventh to the tenth magnitudes, countless tiny star-points of lesser magnitudes, star-clusters of all descriptions, and nearly three hundred nebulæ—all crowded into this seemingly limited space.

CHAPTER V.

THE MILKY WAY.

"Who knoweth not in all these that the hand of the Lord hath wrought this?"—JOB 12:9.
"One star differeth from another star in glory."—1 COR. 15:41.

THE Milky Way forms a soft band of light round the whole heavens. In the southern hemisphere as in the northern hemisphere it is to be seen.

In some parts the band narrows, in some parts it widens. Here it divides into two branches. There we find dark spaces in its midst. One such space in the south is to the naked eye so black and almost starless as to have been named the Coal-sack. But photography shows abundance of stars to lie there also.

All along, over the background of soft dim light, lies a scattering of brighter stars shining on its surface.

Much interest and curiosity have long been felt about this mysterious Milky Way. That it consists of innumerable suns, and that our sun is one among them, has been believed for a considerable time. But other questions arise. How many stars does the Milky Way contain? What is its shape? How far does it reach?

No harm in asking the questions, only we have to be satisfied in astronomy to ask many questions which cannot yet receive answers. No harm for man to learn that the utmost reach of his intellect must fall short in any attempt to sound the depths of the universe—even as the arm of a child would fall short in seeking to sound the depths of the ocean over the side of a little boat.

For attempts have been made to sound the depths of our star-galaxy out of this little earth-boat.

The idea first occurred to the great Herschel, and a grand idea it was—only a hopeless one.

He turned his powerful telescope north, south, east, west. He counted the stars visible at one time in this, in that, in the other directions. He found a marked difference in the numbers. The portion of sky seen through his telescope was about one quarter the size of that covered by the moon. Sometimes he could merely perceive two or three bright points on a black background, at other times the field of his telescope was crowded.

In the fuller portions of the Milky Way he had four or five hundred stars under view at once. In one place he saw about one hundred and sixteen thousand stars pass before him in a single quarter of an hour.

Herschel took it for granted that the stars of the Milky Way, uncountable in numbers, are as a rule much the same in size; so that brightest stars would as a rule be nearest, and dimmest stars would as a rule be farthest off. Where he found stars clustering thickly, beyond his power to penetrate, he believed that the Milky Way reached very far in that direction. Where he found black space, unlighted by stars or lighted by few stars, he decided that he had found the borders of our galaxy in that direction.

Following these rules which he had laid down, he made a sort of rough sketch of what he supposed might be the shape of the Milky Way. He thought it was somewhat flat, extending to a good distance breadthways and a much greater distance lengthways, and he placed our sun not far from the middle. This imagined shape of the Milky Way is called "the Cloven-Disc Theory." To explain the appearance of the Milky Way in the sky Herschel supposed it to be cloven or split through half of its length, with a black space between the two split parts.

It seems that Herschel did not hold strongly to this idea in later years, and the rules on which he formed it no longer hold good.

For how do we know that the stars of the Milky Way are as a rule much the same in size? Certainly the planets of the Solar System are very far from being uniform; and the few stars whose weight can with any certainty be measured seem to vary considerably. There is a great difference also between the large and small suns in many of the double stars.

Again, although bright stars undoubtedly are as a rule the nearest and dim stars farthest off, there are many exceptions. Look at Sirius and 61 Cygni—Sirius, the most radiant star in the heavens, and 61 Cygni, almost invisible to the naked eye. According to this rule, 61 Cygni ought to lie at an enormous distance beyond Sirius. Yet in actual fact Sirius is the farther away of the two.

For aught we know there may be many small telescopic stars much nearer to us than many of the brighter stars which we can plainly see with the naked eye, and only invisible without the help of the telescope because of their smallness, not so much because of their distance.

This line of reasoning shows how easily mistakes may have been made in a matter of no slight importance. Some chapters back I spoke of the light travelling to us from far-distant stars—stars so distant that their light must have taken thousands of years to reach us.

Only there is no *must*. We do not actually know the fact. A writer says: "The illustrious Herschel penetrated on one occasion into this spot"—the star-cluster on the sword-hand in the constellation of Perseus—"until he found himself among depths whose light could not have reached him in less than four thousand years."

So Herschel believed, but he did not know. There is no certain "could not" in the matter. The distance of those stars had not and has not been mathematically measured. Herschel judged of it by their dimness, by the strong power needed to make them visible, and by the rules which he had adopted as most likely true.

This was no sure mode of judging. The stars *might* be as far distant as he supposed. On the other hand they *might* be very much nearer. If too far for our measuring-line to reach them, they may be only just too far; and their seeming smallness and dimness, instead of being caused by vast reaches of space between, may simply be caused by actual smallness or actual dimness. There are little suns as well as great suns in the Milky Way. It is possible that hundreds of little suns may exist for every single specimen of a great sun.

Again, when Herschel found black spaces in the heavens, almost void of stars, and believed that he had reached the outside borders of the Milky Way, he might be right or he might be mistaken. The limit might lie there, or thousands more of small stars might extend in that very direction too far off for their little glimmer to be seen through the most powerful telescope.

A good deal of attention has been paid to the arrangement of stars in the sky. The more the matter is looked into the more it appears that stars are neither regular in size nor regular in distribution. It would seem that they are not flung broadcast through the heavens, each one alone and independent of the rest. They are placed often, as we have already seen, in pairs, in triplets, in quartettes, in clusters. Also the great masses of them in the heavens seem to be more or less arranged in streams and sprays and spirals. It should, however, be understood that what may look to us like streams or spirals may in reality be nothing of the kind. Distance is very deceptive. Even on earth any object far off may appear to be close to another object from which it is really separated by many miles; and if this is so on our little earth, how much more easily may one be mistaken in celestial landscapes.

We must now try to form a few clear ideas about the richer and poorer parts of the heavens as viewed from earth.

In considering stars of the first six magnitudes only—stars visible to the naked eye—a somewhat larger number is found in the southern hemisphere than in the northern hemisphere. In both hemispheres there are regions densely crowded with stars and regions by comparison almost empty.

It has been long questioned whether the number of bright stars is or is not greater in the Milky Way than in other parts of the sky.

Careful calculations have at length been made. It appears that the whole of the Milky Way—that zone of soft light passing round the earth—covers, if we leave out the Coal-sack and other such gaps, between one-tenth and one-eleventh of the whole heavens.

The entire number of naked-eye stars, or stars of the first six magnitudes, does not exceed six thousand, and of these, eleven hundred and fifteen lie scattered along the bed of the Milky Way stream.

If the brighter stars were scattered over all the sky as thickly as throughout the Milky Way, their number would amount to twelve thousand instead of only six thousand. This shows us that the higher magnitude stars really are collected along the Milky Way in greater numbers than elsewhere.

In the dark spaces of the Milky Way, on the contrary, bright stars are so few that if they were scattered in the same manner over all the sky, their present number of six thousand would come down to twelve hundred and forty—which would be a serious loss.

No rest, no quiet, no repose in that great universe which to our dim eyesight looks so fixed and still, but one perpetual rush of moving suns and worlds.

For every star has its own particular motion; every sun is pressing forward in its own appointed path.

And among the myriads of stars — bright blazing spheres of white or golden, red, blue, or green radiance, sweeping with steady rush through space, our sun also hastens onward.

When I speak of the sun's movement, it must of course be understood that the earth and planets all travel with him, much as a great steamer on the sea might drag in his wake a number of little boats. From one of the little boats you could judge of the steamer's motion quite as well as if you were on the steamer itself. Astronomers judge partly of the sun's motion by watching the seeming backward drift of stars to the right and left of him; and the watching can be as well accomplished from earth as from the sun himself.

After all, this mode of judging is and must be very uncertain. Among the millions of stars visible we know the distances, roughly, of less than one hundred; and every star has its own real motion, which has to be separated from the apparent change of position caused by the sun's advance.

It seems now probable that the sun's course is

The Milky Way. 305

directed more or less towards the neighborhood of the constellation Hercules. But if orbits of suns, like orbits of planets, are ellipses, he will curve away sideways long before he reaches Hercules.

Another investigation gives the neighborhood of 61 Cygni as our "next station."

It is difficult to give any clear idea of the immensity of the universe—even of that portion of the universe which lies within reach of our most powerful telescopes. How far beyond such limits it may reach we lose ourselves in imagining.

Earlier in the book we have supposed possible models of the Solar System, bringing down sun and worlds to a small size, yet keeping due proportions. What if we were to attempt to make a reduced model of the universe—that is, of just so much of it as comes within our ken?

Suppose a man were to set himself to form such a model, including every star which has ever been seen.

Let him have one tiny ball for the sun and another tiny ball for Alpha Centauri, and let him as a beginning set the two *one yard apart*. That single yard represents ninety-one millions of miles two hundred and thirty thousand times repeated. Then let him arrange countless multitudes of other tiny balls at due distances—some five times, ten times, twenty times, fifty times as far away from the sun as Alpha Centauri.

It is said that the known universe, made upon a model of these proportions, would be many miles in length and breadth.

But the model would appear fixed as marble. The sizes and distances of the stars being so enormously reduced, their rates of motion would be lessened in proportion. Long intervals of time would need to pass before the faintest motion in one of the millions of tiny balls could become visible to a human eye.

CHAPTER VI.

READING THE LIGHT.

"God said, Let there be light; and there was light." GEN. I : 3.

MANY times in the course of this little book mention has been made of an instrument called the Spectroscope, to which much of our present knowledge of the heavens is due. The subject of Spectrum Analysis is too complex to be fully discussed in a volume of this kind; but a few words of explanation may be desirable.

By means of the spectroscope we know, with almost certainty, many of the substances which are contained in the sun, in the stars, in nebulæ. How large, or how far away, or how quickly moving, some of the heavenly bodies are, could be discovered through the telescope without help from the spectroscope; but the actual *make* of them lay, until recent years, beyond our grasp.

Not upon earth alone, but also in the sun, exist iron and sodium, copper and zinc, magnesium, cobalt, and many other substances with which we are familiar—notably, hydrogen.

Every metal may be either in the solid or in the liquid or in the vapor form. Iron, as we commonly see it, is solid; in other words, it is frozen, like ice. Most metals freeze at a much lower temperature than water does. Just as a certain increase of heat will turn ice into water, so a certain increase of heat will transform solid iron into liquid iron. And just as yet greater heat will turn water into steam, so very much greater heat will transform liquid iron into vapor of iron, or iron gas. A little heat will do for ice what immense heat will do for iron.

The intense heat of the sun causes metals which are solid upon earth to float as glowing vapor in the solar atmosphere. Indeed, that dense and far-reaching atmosphere is largely composed of such metals. Whether iron and other earthly metals can be found anywhere in or near the sun in anything approaching to a solid form is more than doubtful; but in the photosphere they probably do exist in a shape bordering on the liquid form, where glowing clouds appear to float, formed of condensed and radiant metals. This state would answer, not to the liquid waters of a river, but to the condensed waterdroplets of a fog or cloud.

Spectrum Analysis, beside teaching us what metals may be found in the sun, has also a word to say about those which exist in some of the stars. Far distant as those suns of light are, we know that in them too are such earthly materials as iron, sodium, magnesium, and a variety of kindred substances, while in many of them hydrogen predominates.

But how can we know all this? How could the wildest guessing grow to more than a guess and reveal to us the actual presence in even our own particular sun of iron or hydrogen, not to speak of more distant suns?

We know it by means of Spectrum Analysis.

The spectroscope may be looked upon as twin-sister to the telescope. A telescope gathers together widely-scattered rays of light into a spot or focus for our use. The spectroscope separates those rays of light into ribbons, sorts them, and enables us to read in them hidden meanings.

When a ray of light reaches us from the sun, that ray is white; but in the whiteness other hues are hidden. A white ray is composite in form, made up of many lesser rays of divers colors blended together. Newton was the first to discover so much.

If a sunlight ray is allowed to pass through a small round hole in the wall it will fall upon the opposite wall, or upon the ground, in a little round patch of light.

But if a piece of glass cut into the shape of a prism is placed in the path of that sunbeam, the round patch of light vanishes. In its stead appear several bands of soft color, each overlapping and melting into the next. Red, orange, yellow, green, blue, indigo, violet—all these become visible. The arrangement of colors is invariably the same.

Now the prism has done two things. First, it has bent the ray out of a straight course, causing it to fall in a different place from where the round spot lay. Secondly, it has broken up or disintegrated the white ray into those differently-tinted rays of which it was made. And the manner of breaking it up is simply this—that one color is always more or less bent than another color, in passing through a prism, therefore all the different tints fall upon different places.

The whole variegated band, whether of sunlight or of any other light, is known as the "spectrum" of that light; and the breaking up of the ray and searching into its make is called the "analysis" of it. Thus the reading of light—whether earthly light, moonlight, sunlight, or starlight—is included under the head of Spectrum Analysis.

At the so-called "lowest" end of the spectrum lies red, the least bent of all the colored rays. At the so-called "highest" end lies violet, the most bent. All other hues visible to our eyes are placed between. Light beyond the violet and below the red we cannot see. That is by no means to say that it does not exist. We can feel the warmth of rays below the red, although we cannot see them; and photographs are taken by means of rays beyond both the red and violet, albeit they are to us invisible. By means of those faint rays photographs can actually be taken of countless stars which no living man can see through the most powerful telescopes. For the rays are *there*, however faint, and they will slowly impress their image upon prepared sensitive paper long exposed to their dim shining, though human eyes cannot gaze long or steadily enough to detect them.

Suppose now that, for the round hole in the wall, we substitute a very narrow slit. Then letting the sunshine again stream through, we once more place a prism of glass in the path of the ray when it has entered the slit.

Bright bands of color are perceived anew, arranged always in the same order, from red to violet. But the different colors no longer overlap as before; and in addition to the bright hues, a great many slender dark lines or gaps in the coloring are to be seen.

These dark lines in the spectrum of sunlight were for a long while a great perplexity. At first it was thought that they might be caused by something in our own atmosphere; but this notion had to be given up. It became manifest that the lines were somehow connected with the sun, not with the earth. Wherever a ray of sunlight was passed through a narrow slit and analyzed, there they were—always the same in position. More and more of them were found through closer observation, but the old ones did not change.

A ray of moonlight was found to contain exactly the same dark lines as a ray of sunlight—and most naturally! For a ray of moonlight really is nothing more or less than a ray of sunlight reflected from the moon.

But in a ray of starlight the lines were found to be entirely different; and this alone is enough to do away with the idea of the lines being due to something in our air. Starlight and sunlight both alike journey through earth's atmosphere. Every star has its own particular spectrum, sometimes bands of color, like our sun's spectrum, with fine dark lines, only the arrangement of lines is never the same as those of our sun, and it is never the same in any two stars.

These dark lines in sunlight and starlight tell us much ; and briefly, as follows :

If the light of a white-hot metal, either solid or liquid, is allowed to pass through the slit and prism of a spectroscope, bands of bright color are seen, one passing into another, as with the solar spectrum.

But under ordinary conditions of bodies in a *gaseous* state the spectrum does not consist of broad hazy bands of color, but of sharp bright lines.

And—mark this—each particular gas has its own especial lines, always occupying the same position. A chemist knows to a certainty how many lines will be seen, in the case of iron gas or sodium gas or hydrogen gas, and where each line will fall.

But suppose—mark this again—that the light comes either from a solid or a liquid or a highly-compressed gas, and passes on its way through a gas medium at a lower temperature, such as a gas-flame or glowing gas atmosphere? In such a case, when the ray has gone through slit and prism and has spread out its spectrum of color bands, *dark lines* will be seen in those bands. These dark lines are gaps or breaks in the light. They mean the absence of part of the sunlight or starlight. For the gas through which that light has travelled on its way hitherward has absorbed or captured part of the light, and only part has been free to continue its journey.

A ray of light from the sun's photosphere has to pass through the sun's atmosphere of gases; and as it passes, each gas in the solar atmosphere takes possession of some part of it, producing dark lines which are peculiar to that gas. In consequence of this, hundreds of tiny breaks are formed in the solar spectrum where the much-robbed ray falls.

If the light of a sodium-flame is passed through the spectroscope two bright lines are always seen, always in the same position.

But in the sun's colored spectrum those two lines are *dark*. For the sodium in the sun's atmosphere has absorbed or stolen exactly that part of the sunlight which would have given the two *bright* sodium lines.

The spectrum of iron means, not two, but hundreds of bright lines. In the solar spectrum these hundreds of lines are dark, because the iron gas in the atmosphere of the sun has absorbed all that part of the light-ray on its journey from the radiant photosphere.

Therefore, when certain *dark lines* are apparent in the spectrum of the sun or of a star, we know that the metal which always shows exactly corresponding *bright lines* is present in the atmosphere of that sun or star. By the same mode the make of comets, of nebulæ, and of other heavenly bodies can be tested and examined.

This may give just a faint idea of the first principles of the new grand branch of Astronomy included under the head of Spectrum analysis. The science is a science of itself, however, and one full of complexity. Chemistry and astronomy here walk hand in hand.

The spectroscope alone can decide whether any particular star is in fact a sun. If the "rainbow-tinted ribbon" of light is there, crossed by thin dark lines, due to the stellar atmosphere through which each ray has passed after quitting the photosphere, then that star may be counted a true sun. Such stars constitute the majority among the hosts of heaven; yet variety without limit is found. In one star the light from the photosphere is almost lost, almost absorbed by the surrounding atmosphere. In another, the light of the atmosphere so overpowers that of the photosphere that only bright gas-lines are seen in the spectroscope — the continuous color-band being comparatively faint.

Besides telling us of the make of the stars, the spectroscope also speaks to us of their motions. A very tiny shift or movement of the little dark lines or bright lines, to right or left, is sufficient. From this can be calculated with, it is believed, astonishing exactness the rate at which a star is rushing towards us or rushing away from us. For if the star is nearing our earth, the light-waves coming from him are pressed together, so to speak, hurrying one upon another; while if the star is retreating, the light-waves are affected in a reverse manner, delayed or pulled apart, if such a term may be used. By this method* also stars are discovered so dim as to be not only utterly invisible to the naked eye, but undiscoverable by the telescope.

* It was Dr. Huggins who inaugurated this method of research upon the heavenly bodies, the results being communicated to the Royal Society in 1868.

CHAPTER VII.

FURTHER THOUGHTS.

"My right hand hath spanned the heavens."—ISA. 48:13.

EARLY in the year 1892 a new star appeared in the constellation Auriga, winning much attention.

On December 10, 1891, it made its first appearance, suddenly, as a star of the fifth magnitude. Two days earlier no star so bright as even the ninth magnitude had been visible in that part of the heavens. On the 20th of December it reached its brightest. On the 8th of February it was again only a star of the fifth magnitude; and after the 7th of March there were "remarkable swayings to and fro of the intensity of the light, set up probably by commotions attendant on the cause of its outburst."* These lessened, and the star fell steadily to only the eleventh, and then by the 1st of April to the fourteenth, magnitude. At the end of April it was still faintly visible.

Nova Aurigæ, as this brilliant stranger has been named, showed under the spectroscope "a rainbow-tinted ribbon, in which splendid groups

* Dr. Huggins.

of bright lines stood out from a paler background; the red ray of hydrogen, Fraunhofer's C., glowed ... like a danger signal on a dark night; a superb quartette of rays shone in the green; shimmering blue bands and lines drew the eye far up towards the violet; the characteristic blazing spectrum, in fact, of a new star was unmistakably present. The interpretation left no doubt that hydrogen played a large part in the conflagration."*

Nor was this all that the spectroscope had to say!

At the close of the last chapter brief mention was made of the mode in which the coming nearer or passing away of a star may be discovered. With the spectrum of Nova Aurigæ an extraordinary thing was seen. There were distinct and unmistakable signs, not of approach alone, or of a receding movement alone, but of both simultaneously. At one and the self-same moment the Nova appeared to be speeding away from us at a rate of two or three hundred miles per second and rushing towards us at much the same rate, the sum of the two motions together amounting to no less than about 550 miles per second!

* Miss Clarke.

At first this was explained by a conjecture that Nova Aurigæ might consist of *two* stars: one of the two hurrying towards us and the other hurrying away from us. But the two-star explanation is already being given up, and astronomers are inclined to think that the rush to and fro is probably only of heated gases on the turbulent surface of the star.

For Nova Aurigæ, like many or most other stars, is most likely a mass of intensely heated and glowing gases, forcibly compressed within by gravitation and spreading outside far away into a furiously-excited incandescent atmosphere.

In any case, the history of the Nova is a remarkable one, and apparently we have not yet come to the end of it. After steadily waning to a very faint star, it has begun again to increase in brightness. What course it will follow in the future, time alone can show.

Mention has been more than once made, in this revised edition, of the probable presence and absence of atmospheres on different heavenly bodies, and of apparent reasons for the same.

An atmosphere is formed of gas-particles, or gas-molecules, in rapid motion. There are different gases, and the particles of one gas differ in speed from those of another gas. Moreover, the colder any gas is, the more slowly its particles move; the hotter any gas, the more vehemently its tiny molecules dash to and fro, incessantly striking against other molecules and striving to escape.

If this is the case with our atmosphere, why should not all the particles of air escape in the course of time, passing into distant space and forsaking us utterly?

So unquestionably they would do but for the restraining force of gravitation. Struggle as they may to get away, earth's strong attraction holds them in leash, and they cannot flee!

Our atmosphere consists mainly of oxygen and nitrogen; and the oxygen gas is essential for animal life. No free hydrogen gas floats in our atmosphere, although enormous quantities of it are present in the sun. It is very possible that once upon a time earth possessed hydrogen gas also in her air, and that she has lost it since. For the particles of hydrogen are exceedingly rapid in their motions, far more so than the particles of oxygen or nitrogen; therefore they need a much stronger restraint to hold them in. And while earth is large and dense enough to keep her nitrogen and oxygen gases prisoners, she has not, it would seem, sufficient attractive power to hold in hydrogen.

If Mars has an atmosphere, or Mercury, we may say with tolerable confidence that no free hydrogen exists in it; because neither Mars nor Mercury could be expected to hold fast the rapidly journeying particles of hydrogen gas. It has even been calculated that, very probably, Mars is somewhere about the smallest world which can permanently retain an atmosphere.

We have already seen that the moon, so far as we can discover, has no atmosphere. If ever the moon did possess one, it has doubtless all wandered away, because her attractive power was too weak to keep it.

This is not merely a guess. The speed of different gas particles is accurately known, and the strength of gravitation needed to control them is known also. It is no difficult matter to find out about how heavy a world must be to overcome by attraction the impetus of the busy little gas-molecules. But while the moon, perhaps, could not keep any atmosphere, and while the earth might be unable to hold in bondage an atmosphere of hydrogen gas, the sun has no such difficulty. His enormous bulk gives such enormous attractive power that the hydrogen in his atmosphere, though violently heated and furiously excited, is still under control.

A most natural, though wide, step carries us from the restless gas-particles, breaking loose from control, to the larger restless heavenly bodies, which appear to do precisely the same thing.

In our Solar System we have worlds and satellites, comets and meteors, under the sun's command, obedient to his attractive power. We have also wandering comets, occasional and passing visitants, travelling at so rapid a pace that all the great attraction of the sun fails utterly to hold them in; and they plunge into the depths of space never to return.

In our Stellar System we have countless suns, double stars, clusters, and perhaps nebulæ, subject one to another, bound together by the bonds of mutual gravitation. But also we have, possibly, Runaway Stars, wanderers from the depths beyond, passing through our Starry System at a headlong rush which, it may be, not all the combined control of all the stars of light can check or hinder. If so, they doubtless will plunge into profounder distances, beyond the utmost stretch of human imagination, never to return.

Verily, the great things and the little things of creation are subject to the same laws!

A few closing words as to the possible comparative ages of different heavenly bodies may be of interest.

The ages of worlds and suns cannot indeed be reckoned by years or by centuries. No record gives us the date, in past millenniums, of the birth of this or that bright orb. Yet, looking upon them from our little world, we see signs which seem to tell, in one of comparative youth, in another of comparative age. Nay, so marked are these signs, and so evidently is one stage succeeded by another stage, that, as in looking on boy and man we may say with confidence that the one has been, or the one will be, just like the other, so in looking on diverse heavenly bodies we may conjecture that as one is another has been or will be.

Positive assertions are indeed, here as elsewhere, out of place. Watching the heavens as we do with limited sight, and for a very brief span, it behooves us to be careful even in conjecture.

Though the marvel is that we can know so much, though the question may well be asked, "What then must the true nature of that MIND be which thus freely associates itself with the things beyond the stars? surely it must find its natural home among the infinite, the infinitely remote both in space and in time!"* still caution linked with humility must be our watchword. For the utmost that we know is as nothing compared with the vast regions beyond which we do not know.

Still the more yonder heavens are searched, the more it does appear that development takes place among those shining hosts, even as among the little existences on earth.

Our moon, for instance, seems to be a very old and used-up world—old not so much in centuries or millenniums as in having reached a certain decrepid and worn-out phase of existence. One man is old and another is young at sixty. In actual decades of centuries the moon may reckon no more recurring birthdays than earth or sun; but in the social standing of the skies she is, or seems to be, very elderly, very much paid out, almost a dead world.

* Dr. Pritchard: "History of Research in Stellar Parallax."

Once upon a time, doubtless, the moon was a shining mass of rotating gases, vague and nebulous in form, and that was her extreme youth—her infancy, so to speak. For her actual birth one would have to go farther back still. Then through ages, as gravitation drew the masses of gas nearer, she probably grew into a tiny sun or star, with a radiant photosphere and an atmosphere of heated gases in commotion. So tiny a sun would cool quickly; and by-and-by the radiance would die out, as the little orb passed from vigorous youth to early middle life, reaching the present heated and molten but not luminous condition of Jupiter. After which would follow the advanced middle-age, such as our earth has now attained to; though, if the moon had by that time lost her atmosphere, it could not well be an inhabited stage. Lastly she has gained her present half-dead phase of extreme old age.

The sun is young still, in the very heyday of his existence. Though now a huge shapely ball, with a photosphere of radiant clouds and an envelope of heated gases, he was, probably, once upon a time, a vast nebula of revolving gases, like such nebulæ as we see now in the skies. A mutual attraction drew the gas-particles slowly together, his enormous hazy bulk gradually lessened and took a definite form. But still he continues and will steadily continue to lessen in size; and this shrinkage is conjectured to cause the heat from which springs his glowing radiance. By-and-by, after countless ages, he may be expected to cool down further yet, first to a molten and then to a solid body, no longer giving out light or heat. Jupiter has passed from the gaseous and radiant to the heated but non-luminous condition. Earth, being smaller, has cooled and shrunk more rapidly than Jupiter, though by no means so rapidly as the still smaller moon.

Thus in Sun, Jupiter, Earth, Moon we seem to see four distinct stages of star-life, the radiant vigor of youth, the calmer warmth of manhood, the composed usefulness of middle life, the used-up chilliness of extreme old age.

This may have been—and, seemingly to us, it has been—the mode in which our God gradually, stage by stage, "made the stars" and " framed the worlds" which float in the universe.

These few paces have been trodden. Step by step we have advanced, from the first early notion of our earth as the mighty centre of the universe to a stage where earth and planets vanish from sight and the great central sun of our system himself shines forth but as one glimmering point of light amid a million stars.

Well may we say, in the words of the inspired writer, as we gaze on the wonders of God's glorious creation :

"When I consider Thy heavens, the work of Thy fingers, the moon and the stars which Thou hast ordained, what is man that Thou art mindful of him? and the son of man that Thou visitest him?"*

And yet there is another side to the matter. Infinite in power, God is also infinite in love. Mighty in the creation of his vast and blazing suns, he is no less mighty in the creation of a blade of grass. The rushing stars of a universe are in his hand, but not a sparrow on earth may fall to the ground without his knowledge. He guides each distant sun in its pathway through the sky, and also he looks from his throne of light to "ponder" the ways of man. He tends His countless furnaces of fire and flame; yet "the very hairs of your head are all numbered."

Let us look into the matter as we will—let us weigh, measure, calculate—let us find our earth to

---

* Psa. 8 : 3, 4.

be but as a grain of fine dust, lost amid myriads of worlds and suns. Still, at the close of all we stand face to face with the simple historical fact that the King of heaven, the Creator of the universe, Himself lived as man for thirty-three years upon earth, died upon earth, rose from death upon earth, and went up again from earth to heaven. That sheds a distinguishing radiance upon our earth which, it may well be, no other world in all the universe can rival.

"Thou art worthy, O Lord, to receive glory and honor and power: for Thou hast created all things, and for Thy pleasure they are and were created." Rev. 4:11.

TABLE OF SUBJECTS.

PART I.

CHAP. I. What is the Earth? Olden beliefs. The Universe. Space. What are Stars? Movements of Earth. Seeming Movements of Stars. The Evening Star. Planets.... 17—27

CHAP. II. The size of the Sun. The distance of the Sun from Earth. Size of Sun and Earth compared. Spots upon the Sun. Photosphere. Prominences. Storms on the Sun. Surface of the Sun-- 28—40

CHAP. III. The law of Attraction or Gravitation. A law of Motion. How both laws work together. What if the Earth went faster? Two Motions of the Earth. Illustration with orange and knitting-needle. Day and Night. The Seasons. Speed of the Earth's surface at the Equator---------------------- 41—54

CHAP. IV. Superior and Inferior Planets. Pathways of the Planets. Two Groups of Planets. Names of chief Planets. Mercury: his size, distance from sun, speed, and length of year. Venus: her size, distance from sun, speed, length of year, appearance as seen from earth, and phases. Mars: his size, distance from sun, length of day and year, speed and appearance. Moons of Mars. The Planetoids: their number, size, length of years, zone, and orbits--- 55—64

CHAP. V. Jupiter: his size, distance from sun, speed, cloudy covering, length of day and year. Four Moons of Jupiter. Saturn: his distance from sun, size, length of day and year, motions, and cloudy covering. Moons and Rings of Saturn. Uranus: his distance from sun, size, speed, and length of year. Moons of Uranus. Neptune: his distance from sun, size, speed, length of year. Moon. Appearance of sun seen from Neptune. What is meant by "a million miles." Size and distance. Supposed Models of Solar System_____ 65—75

CHAP. VI. Journey to the Moon. Distance of the Moon from Earth. Size of the Moon. Two ways of thinking about the Moon. Motions of the Moon. Only one side seen from Earth. Moonland. Supposed Scenery. Supposed view of the heavens. Moon-mountains and plains. Craters. Shadows. Sunset and night in the Moon. Intense cold_____ 76—87

CHAP. VII. Comets. Old dread of a collision. Nucleus, coma, and tail. Lightness of a Comet. Number of Comets known. Numbers of Comets supposed to exist. Comet-visitors. Pathways of Comets. Singular nature of Comets. Changes seen to take place in Comets. Halley's Comet. Encke's Comet. Newton's Comet_____ 88—95

CHAP. VIII. Shooting Stars. A cannonade. Numbers of meteors seen nightly. Numbers supposed to fall daily to earth. What becomes of the Shooting-stars. Meteorite-dust. Aerolites. Meteors or fire-balls. Numbers of Meteor-systems met by earth. Numbers supposed to exist. August and November Rings. Grand displays every thirty-three years. Saturn's Rings. Zodiacal Light_____ 96—104

CHAP. IX. Away from the Solar System. The Stars. How many stars are there? Star-magnitudes. Distances of Stars. Measurement of distance. Distance of Alpha Centauri. Speed of Light. Light-journeyings from sun and stars---------- 105—117

CHAP. X. Apparent movements of stars. Real movements of stars. Capella, Sirius, etc. Runaway stars. Arcturus: rate of speed. Journey of the sun. Where is he going?-------------- 118—131

PART II.

CHAP. I. What the System contains. Is there Life on the Planets? Sun-rays and their work. Three Movements of the Earth. Journey of Sun through Space. Earth's orbit. Planets' orbits. How to draw a circle and an ellipse. Comets' orbits. Family influences. "Great" and "little."-------------------- 135—148

CHAP. II. Sun-spots. Umbra, Penumbra, and Nucleus. Size of Spots. Possible explanation of Spots. Real and seeming changes in Spots. Sun-cyclones. Solar-outbursts and Magnets of Earth. "Willow-leaves." Eclipse of the Sun. Partial, Total, and Annular Eclipses. Eclipse of 1860--------------- 149—161

CHAP. III. Storms seen at edge of Sun. Great solar outburst. Speed of hydrogen-cloud fragments. Red prominences, their shape and height. Corona described. Bulk and weight of Sun. Weight on surface of Sun. How far Attraction of Sun reaches -- 162—171

CHAP. IV. Is there Life on the Moon? Size of Moon. Absence of Atmosphere. Reflected sunlight and earth-shine. Phases of Moon described. Illustration of Phases. Eclipse of the Moon. Total and Partial Eclipses---------------------------- 172—181

CHAP. V. Two ways of thinking about the Moon. Orbit of moon. Moon's motions governed by the Sun. Perturbations caused by

Earth. Attraction of Moon and Ocean-tides. Unknown side of moon. No water in moon. Craters. Sunset-lights. Falling meteorites. Names of some chief Craters and Mountain-ranges -- 182—193

CHAP. VI. Mercury: his speed, length of day, slope of axis, orbit, varying distance from sun. Sun as seen from Mercury. Dense Atmosphere. Life on Mercury or no? Weight at surface of Mercury. Venus: her orbit. Transits of Mercury and Venus. Distance of Venus from Earth. Appearance and Phases of Venus. Appearance of Earth, seen from Venus. Length of day, climate, slanting axis of Venus. Life on Venus? Mars: his phases, moons, size, density, atmosphere, weather, geography, and distance from Earth. Water and Snow. Map of Mars. --- 194—208

CHAP. VII. Jupiter: his size, weight, shape, slope of axis, turning on axis, seasons. Sun as seen from Jupiter. Jupiter seen through a telescope. Dark and bright Belts of Jupiter. Poles of Planets. Northern and southern heavens. West to east motions of planets. Atmosphere of Jupiter. Possible explanation of Belts. Bright white spots. Motions of Jupiter's Bands. Winds and Hurricanes. Supposed present state of Jupiter. Life on Jupiter? Jupiter's moons: their size, distance from Jupiter, eclipses, names------------------------------ 209—218

CHAP. VIII. Saturn: his size, weight, cloudy atmosphere, supposed present condition. Moons of Saturn. Square-shouldered appearance of Saturn. The Rings: their number, color, breadth. Shining of Rings by reflected light. Winter on Saturn. Appearance of Saturn-------------------------------------- 219—226

CHAP. IX. Distance of Uranus from Saturn and from Neptune. Discovery of Uranus. Orbits of planets. Discovery of Neptune.

Uranus and Neptune : their size, density, length of year, moons. Sun as seen from Uranus and Neptune................ 227—235

CHAP. X. Connection between Comets and Meteors. Meteor Rings. November System: supposed weight of Meteors. Grand display of 1866. Orbit of November System. Chief cluster: its supposed length, depth, and thickness. How many Meteor-Rings in the Solar System ?.............. 236—243

CHAP. XI. Comet of 1811. Halley's Comet. Comet of 1843. Donati's Comet. Texture of a comet's tail. Weight of Comets. Changes seen in Comets. Motions of Comets. What are Comets and Meteorites made of? Meteorite dust. Suppositions. Comet-visitors, and where they come from...... 244—253

PART III.

CHAP. I. Space beyond Solar System. Our Sun a Star. Speed of Light. Immense Distance of Stars. Distance of Alpha Centauri. Sizes of Stars. Four chief orders of Stars. What Stars really are. Movements of Stars..................... 257—265

CHAP. II. 61 Cygni: double star, his speed, distance, weight. Alpha Centauri: double star, his position in heavens, speed, distance, weight. Sirius: his color, distance, size, motions, satellite. What Sirius is made of..................... 266—277

CHAP. III. Variable Stars. Mira, Algol, Betelgeuse, Delta Cephei, Sun. Temporary Star of 1572. Two kinds of Double Stars. Quadruple Star near Vega. Stars of many colors. Examples of white, yellow, and red stars. Group of colored stars in southern heavens................................. 278—286

CHAP. IV. The Milky Way. Globular Clusters. Colored Clusters. What are the Clusters? Worlds without night. Nebulæ. What are the Nebulæ? Nebulæ in Orion. Different kinds of Nebulæ. Magellanic Clouds------------------------- 287—295

CHAP. V. Appearance of the Milky Way. What the Milky Way consists of. Sounding the Universe. The Cloven-Disc Theory. Ideas past and present. Arrangement of Stars in the sky. Motions of Stars. Motion of the Solar System. Supposed Model of the Universe------------------------------ 296—306

CHAP. VI. Spectrum Analysis. Metals: solid, liquid, and vapor forms. Metals in Sun and Stars. Use of a Prism. A broken-up ray of sunlight. Bands and Lines. Star-motions-- 307—316

CHAP. VII. Nova Aurigæ. Atmospheres of heavenly bodies. Runaway Comets and Stars. Comparative ages of Sun, Jupiter, Earth, and Moon. Conclusion--------------------- 317—328

Books by Agnes Giberne.

THE ANDERSON'S; Brother and Sister.
A story. 347 pages. 12mo. 10 illus. $1 25.

"The heroine is a fine character, depicted with much skifulness of touch, and the moral teaching of the book is of the purest kind." *New York Observer.*

THE DALRYMPLES.
A story. 12mo. $1 25.

BESIDE THE WATERS OF COMFORT.
Square. Gilt edges. $1 25.

"This compilation was begun with no idea of publication, simply to meet the compiler's own need in a time of sorrow."

SUN, MOON, AND STARS.
A book of astronomy for beginners. 334 pages. 12mo. Illustrated. $1 25.

AMONG THE STARS; or Wonderful Things in the Sky.
321 pages. 12mo. Illustrated. $1 25.

THE WORLD'S FOUNDATIONS; or Geology for Beginners.
334 pages. 12mo. Illustrated. $1 25.

FATHER ALDUR.
A water story. 376 pages. 12mo. Illustrated. $1 25.

THE OCEAN OF AIR.
398 pages. 12mo. Illustrated. $1 25.

AMERICAN TRACT SOCIETY,

Helps to Bible Study.

Dictionary of the Holy Bible.

By the Rev. W. W. Rand, D. D. 720 pages. 8vo. Cloth, $2. Sheep, $2 50. Mor., $3 50.

Revised in the light of recent researches in Bible lands, and enlarged from the popular edition of which over 200,000 copies have been sold. It is printed from new type.

"Beyond all question the best Bible Dictionary that is before the public." *Presbyterian Review.*

"The most valuable book the American Tract Society ever published." *Rev. Dr. Hamlin.*

Bible Text-Book.

12mo. 232 pages. Cloth. 90 cents.

"This is a short yet very comprehensive cyclopædia of the contents of the Holy Scriptures; in fact, a '*Topical* Concordance.' All Bible places, persons, and subjects are arranged alphabetically, and under each word are given the texts bearing upon the same."

Cruden's Concordance.

Cloth. $1. Roan, sprinkled edge. $2 25.

Edited by Rev. John Eadie. This book, containing over 150,000 distinct references to passages in the Scriptures, is invaluable to the student of the Word. It is printed in plain, clear type on fine paper, making an elegant octavo volume of 561 pages.

Biblical History and Geography.

Rev. H. S. Osborn, LL. D. New edition, with Index. Large 12mo. 312 pages. $1 25.

Bible Atlas.

A series of new and beautiful maps made by Johnston, of London, the letterings of which are in unusually large and distinct type. Paper, 25 cts.

AMERICAN TRACT SOCIETY,

www.ingramcontent.com/pod-product-compliance
Lightning Source LLC
Chambersburg PA
CBHW020228240426
43672CB00006B/457